# Advances in Experimental Medicine and Biology

Volume 1069

More information about this series at http://www.springer.com/series/5584

María Elena Álvarez-Buylla Roces
Juan Carlos Martínez-García
José Dávila-Velderrain
Elisa Domínguez-Hüttinger
Mariana Esther Martínez-Sánchez

# Modeling Methods for Medical Systems Biology

Regulatory Dynamics Underlying the
Emergence of Disease Processes

 Springer

María Elena Álvarez-Buylla Roces
Ciudad Universitaria, UNAM
Instituto de Ecología
Ciudad de México
Distrito Federal, Mexico

José Dávila-Velderrain
CSIL
Massachusetts Institute of Technology
Cambridge, MA, USA

Mariana Esther Martínez-Sánchez
Ciudad Universitaria, UNAM
Instituto de Ecología
Ciudad de México
Distrito Federal, Mexico

Juan Carlos Martínez-García
Departamento de Control Automático
CINVESTAV
Ciudad de México
Distrito Federal, Mexico

Elisa Domínguez-Hüttinger
Centro de Ciencias Matemáticas
UNAM
Morelia, México

ISSN 0065-2598          ISSN 2214-8019   (electronic)
Advances in Experimental Medicine and Biology
ISBN 978-3-030-07748-8          ISBN 978-3-319-89354-9   (eBook)
https://doi.org/10.1007/978-3-319-89354-9

Printed on acid-free paper

This Springer imprint is published by the registered company Springer International Publishing AG part
of Springer Nature.
The registered company address is: Gewerbestrasse 11, 6330 Cham, Switzerland

*To the memory of my beloved father, Ramón Álvarez-Buylla, and as a tribute to my mother, Elena Roces, whom I love and admire immensely. Together they gave me the gift of life and showed me the path to an integrative scientific agenda.*

*María Elena Álvarez-Buylla Roces*

*With immense gratitude to the memory of my beloved parents, Julia García Peña and Carlos Martínez Rodríguez. I will always remember them, as I will always remember my sisters Imelda Soledad and Luz María.*

*Juan Carlos Martínez García*

# Preface

The motivation of this volume stems from our experience on plant development using bottom-up system biology approaches, as well as our investigations in several human health processes. We aim at integrating both types of research experiences in the hope of contributing to understanding the mechanisms involved in the emergence and progression of complex degenerative diseases, such as cancer, diabetes, and atopic dermatitis, among others. These diseases constitute a significant health problem, consuming considerable parts of the national budgets for health care in both rich, developed countries and in poorer developing countries.

Shortly after the Human Genome Project, several causal links could be established between individual genetic components (mutations, polymorphisms) and diseases with a simple genetic basis. These stories of success opened the promise that more complex human health conditions, such as chronic degenerative diseases, could be understood, prevented, and cured with similar approaches. Nonetheless, as high-throughput data accumulate at the genomic, transcriptomic, proteomic, or even metabolomic levels, it has become clear that systems-level approaches are required to complete this task, and that these efforts need to go beyond the simple enumeration of parts, and rather should aim at attaining a more profound understanding of the structural and functional characteristics of the systems involved, including details about the underlying interactions. Such systems imply many components and also nonlinear interactions (for example, positive and negative feedback loops) and are, therefore, considered to be complex. Further, these nonlinear interactions among network components are responsible for the emergence of healthy or disease characteristics and behaviors at different levels of organization (from genetic and non-genetic molecular components to networks, from these to aggregates of cellular networks and tissue patterns, and from these to three-dimensional structures, organs, or functional systems or whole organisms, in interaction with the environment) that cannot be explained on the basis of detailed functional studies of the isolated components.

Such genetic approaches and molecular studies, however, in conjunction with epidemiological data, have been useful to show that a large proportion of the cases of complex diseases are modulated or elicited by a combination of environmental and genetic factors, rather than by single genetic causes. Furthermore, both types of causal factors should be studied in the context of the complex systems in which they participate. It remains a challenge to pinpoint the exact contribution of environmental and genetic factors in the emergence of complex diseases, to characterize their early and asymptomatic stages for a better prognosis and prevention, and to design treatment regimens that successfully revert the symptoms using different combinations and doses of the available drugs. Despite this complexity, some large-scale studies are starting to suggest that chronic degenerative diseases may be modulated by the lifestyle of the individual. Thus, while genetic variation among individuals can affect susceptibility or propensity to develop complex disease conditions, these genetic factors are not sufficient to trigger the onset of pathogenic processes. It remains to be elucidated how exactly so-called lifestyle factors, such as diet, smoking or drinking habits, and exercise habits, among others, affect disease emergence, progression, prevention, and reversion.

Up to now, mainly statistical analyses have been used to establish correlations between environmental lifestyle factors and complex chronic degenerative diseases. However, systems biology approaches are required to understand the mechanisms underlying the correlations or statistical trends between lifestyle and complex disease prevalence, as well as to contribute to a scientific approach to complex disease prevention and treatment. These allow the analysis of complex diseases from an integrative, quantitative, and dynamic perspective. The basic building block of such systems biology approaches are dynamical mathematical models of the regulatory complex networks of genetic and non-genetic components underlying complex diseases which are iteratively constructed and validated with experimental and clinical data. Eventually, such models can be developed for all the implied levels of organization, including the environmental factors involved and even the socioeconomic conditions of people. This is an enormous task that we do not pretend to cover with the volume at hand. Our aim is much more humble and limited, although we consider the approach summarized here to be a starting point towards this larger endeavor.

With this volume we propose a systems biology approach to biomedicine that implies integrating well-curated functional data on molecular genetic components and interactions, mainly of transcriptional factors and signal transduction pathways, into dynamical regulatory networks. We present the basic tools, concepts, and methodologies that are used to construct, validate and analyze these mathematical models, and illustrate our approach with three biomedical examples. We envision this approach as a first necessary building block of a larger and multi-level endeavor in systems biology biomedicine.

The approach put forward here constitutes a first step to explore, with a rigorous and formal basis that explicitly considers a complex systems approach, why and how the modulation of the environment or, more broadly, the lifestyle of an individual, may impact the emergence and progression, and eventually the cure, of complex diseases.

We sustain that the approach that we propose is a necessary first step to envision novel approaches in health care policy that may aid diminishing the budgetary burden implied by the ever-increasing prevalence of these types of human health conditions. Also, the proposed approach may help determine the nature of the socioeconomic structural basis of the human suffering implied in the increase of these types of diseases in both underdeveloped poor countries and more developed and rich ones, despite the tremendous technological advances in biomedicine.

Distrito Federal, Mexico                     María Elena Álvarez-Buylla Roces
Distrito Federal, Mexico                         Juan Carlos Martínez-García
Cambridge, MA, USA                                   José Dávila-Velderrain
Morelia, Mexico                                  Elisa Domínguez-Hüttinger
Distrito Federal, Mexico                   Mariana Esther Martínez-Sánchez
February 2018

# Acknowledgements

All work and achievements are collective. This book is not an exception; it is coauthored by five authors. Nonetheless, many more students, teachers, postdoctoral fellows, and collaborators have contributed with their work, ideas, and their enthusiasm to what we have been able to synthesize in this volume. We thank them all heartedly. We particularly mention and thank Carlos Espinosa-Soto and Luis Mendoza for their initial contributions in Dr. Alvarez-Buylla's laboratory, both as undergraduate and graduate students. Their insight, technical abilities in dealing with network modeling, and hard work greatly contributed to the development of the systems biology approach that we have summarized in this volume. Among our collaborators, Drs. Pablo Padilla and Carlos Villareal have been particularly important in helping to formalize our proposals. Dr. Rafael Barrio has been a very important leader and collaborator toward the multi-level models that we have started to explore in flower and root morphogenesis and will likely inspire new ones to come. We also acknowledge the support from our institutions, the National Autonomous University of Mexico (UNAM), the Mexican National Science Council (CONACYT), and the Center for Research and Advanced Studies of the National Polytechnic Institute of Mexico (Cinvestav-IPN).

# Introduction

During the last decades, data on the molecular genetic components implied in human-health and human-disease have accumulated. This trend has posed the challenge of creating a systemic approach to integrate and analyze such data, and to incorporate non-genetic components to further understand the mechanisms involved in the transition from healthy to pre-clinical and eventually to diseased human conditions. We are particularly interested in contributing to understanding and preventing complex chronic degenerative diseases such as cancer. These diseases remain a public health challenge even in developed, rich countries.

Our aim in this volume is to propose and exemplify a bottom-up approach for biomedical systems biology. We focus on a network-based modeling approach to integrate data on genetic and non-genetic data on the complex regulatory interactions that underlie the transitions of cell states involved in the emergence and progression of complex diseases. We and other researchers have claimed that such diseases emerge from the same systems-level mechanisms that underlie normal developmental processes (i.e., cell differentiation and morphogenesis). Such systemic mechanisms involve, as a first building block, complex intracellular regulatory networks and signal transduction mechanisms that mediate cell response to microenvironmental factors.

We have organized this book in three chapters. In the first one, we summarize the conceptual framework that guides us. In order to understand the emergence and progression of complex human diseases, it is crucial to understand how genes map unto such phenotypic states or conditions. Such mapping necessarily implies the existence of developmental mechanisms underlying cell differentiation and those that rule how such cells organize in space and time to form tissues and three-dimensional organs or structures during morphogenesis. To understand how these processes are altered in the transition from healthy to preclinical and to diseased states, rather than focusing on the role of isolated components, we are interested in a systems biology approach. Top-down systemic approaches have focused on gathering information on as many components as possible to infer the structure and function of the systems involved in the disease emergence and progression based on several sophisticated statistical and large data-set analyses. In contrast, we focus on integrating small well-curated are complex intracellular networks that been involved in the emergence and progression of chronic degenerative diseases.

In this volume, focus on a system level bottom-up approach, where we aim to determine, which sets of components and to those that have been empirically shown to correspond to:

- healthy,
- preclinical,
- and diseased conditions.

So our aim is to uncover relatively small although complex networks that map geno-types unto a space of phenotypical states (healthy, preclinical, and disease). Such networks are studied using dynamical system models grounded on well-curated functional data. Such networks also enable us to study the critical restrictions for recovering time-ordered transition patterns from one state to another, and to systematically assess the role of perturbations, such as genetic and environmental risk factors, but also different treatments, on the progression, reversion, and prevention of the disease process. Our systemic approach to study several aspects of plant development, as well as some specific medical conditions, serves as a methodological and theoretical basis to apply our bottom-up systems-level modeling approach to study the emergence and progression of complex diseases presented in this volume.

In the first chapter, we use a non-formal language to summarize the key concepts and approaches to be developed in the rest of the book. We identify what we refer to as "core regulatory network modules", which correspond to particularly thoroughly characterized sets of components and interactions in particular cell differentiation and/or morphogenetic processes. Several of such processes are altered during the emergence and progression of particular diseases. We aim at contributing to understand such transitions, and how they are modulated by genetic and environmental risk factors or different treatments to converge or deviate from pathological states.

Based on our own and other researchers' work we argue that a combination of Boolean network modeling with differential equations, the latter ones used in cases where experimental data is available to calibrate quantitative models, can be used in biomedical bottom-up studies to answer several clinically relevant questions, such as:

- What is the role of individual genetic or environmental risk factors, alone or in combination, in triggering the onset of disease?
- What is the most effective strategy to prevent disease progression in high-risk patient cohorts?
- What is the minimal treatment dose that effectively reverses the disease?

The important and constructive role of stochastic fluctuations in the context of complex nonlinear systems underlying cell differentiation and morphogenesis under both healthy and disease conditions is also discussed. To this end, we explain different formalization and quantification approaches to address the transitions among cell types in the context of Waddington's epigenetic landscape. Hence, relevant concepts on deterministic as well as stochastic modeling are discussed and we exemplify how single-cell models can be expanded to models of populations of cells attaining different fates independently of each other, or in a concerted manner once cell–cell communication mechanisms are also considered. Likewise, we discuss phenotypic plasticity and cellular reprogramming, which are crucial in several complex biomedical conditions. Furthermore, we emphasize the key role of feedback mechanisms both in intracellular networks and between intra and extracellular components and mechanisms, as well as the richness of dynamical behaviors that emerge from the coupling of processes that occur at different time-scales.

The technical, computational, and mathematical techniques and concepts are formalized and detailed in the second chapter of this volume. For this, we provide graphical as well as relatively simple examples to clarify the tools described. We also provide sources of code and actual programming resources that the reader can access to be able to use the proposed modeling approach to other examples or conditions under study. The focus is on what we call a first building block of a modeling bottom-up approach to (clinical) biomedical systems biology: the integration of conserved or robust gene regulatory network modules, mainly composed by transcriptional regulators, and their link to key signal transduction pathways that dynamically connect such intracellular modules to microenvironmental conditions.

Finally, in the third chapter we present three examples of systems that are involved in different types of human complex diseases or health conditions. We first present our modeling contributions to integrated data on molecular components (from processes such as epithelial and mesenchymal cell differentiation, inflammation, and cell-cycle regulation) that have been associated to the emergence and progression of epithelial cancer and in vitro spontaneous immortalization that imply the transition from normal epithelial cells to mesenchymal cells. In this case, we present a single-cell regulatory network model in its discrete and continuous specification and incorporate epigenetic landscape modeling to test if the proposed core regulatory network module is not only important to understand how the different cell types involved emerge (normal epithelial cells, altered cells under chronic inflammation and mesenchymal cells), but if the restrictions proposed in the regulatory network model under study also underlie the most probable time-ordered cellular transition pattern, that has been observed in vitro and in vivo during epithelial cancer progression: normal epithelial cells, altered cells after chronic inflammation, and mesenchymal cells.

As a second example, we put forward a network model that has shown to be an important component of the systemic mechanisms implied in one of the main cellular processes of the immune system: differentiation and plasticity of CD4+T cells that underlie chronic inflammation. The proposed module includes transcription factors, but also signaling molecules and cytokines. In this case we analyze and experimentally validate, using both normal and altered genetic and non-genetic conditions, the proposed Boolean network model. Also, we approximate the discrete model to a continuous model that enables us to address how changes in the decay rates of each one of the components reshapes the epigenetic landscape (multidimensional quasi-potential that restricts the transitions from one network stable state to others) that emerges from the proposed network, and hence we are able to address which nodes within the proposed networks are more important to explain or predict state transitions in the system under analyses. With this kind of extension of the networks models, we are able to analyze how normal CD4+T cell differentiation and plasticity may be altered under different health conditions. In particular, we have explored the role of inflammatory molecules (i.e., cytokines) and hyperinsulinemia conditions.

Finally, in the last example, we present a model to study the onset and progression of *atopic dermatitis*. In this case, we use with time-scale separation models and to explore how disease emerges from dysregulation of the complex interplay between intracellular mechanisms that shape phenotypic decisions at the cellular level tissue-level processes controlling epidermal homeostasis, immune responses, and infection. Specifically, we systematically assess how different genetic (mutations, polymorphisms) and environmental (pollution, pathogen load) risk factors, alone or in combination, affect the onset and then the progression of the disease. Also, we show how such a model can be used to devise effective treatment strategies that can prevent or revert the disease dynamics, using minimal doses and durations of drugs. We also illustrate how the feedback with clinicians and the validation of the model with clinical data significantly enriches the bottom-up system biology approach that we propose to biomedicine.

In the closing section of this book, we enlist several challenges and important next steps that include multilevel frameworks with explicit spatial modeling, as well as the importance of hybrid models that have been proposed before in other studies. Such modeling approaches will enable more detailed considerations of systems-level mechanisms involved in the tissue and organ levels of organization involved in phenotypic transitions from healthy to pre-clinical and ill conditions. In the present volume we have focused on the first building block, which implies the assemblage of complex regulatory intracellular networks formed by transcription factors and their feedback interactions with signal transduction pathways that respond to microenvironmental factors and tissue-level processes.

# Contents

# List of Figures

# Chapter 1
# Medical Systems Biology

## 1.1 Introduction

The aim of this volume is to encourage the use of systems-level methodologies to contribute to the improvement of human health. We intend to motivate biomedical researchers to complement their current theoretical and empirical practice with up-to-date systems biology conceptual approaches. Our perspective is based on deep understanding of the key biomolecular regulatory mechanisms that underlie health, as well as the emergence and progression of human disease. We strongly believe that the contemporary systems biology perspective opens the door to the effective development of novel methodologies to prevention. This requires a deeper and integrative understanding of the involved underlying systems-level mechanisms. To explain our proposal in a simple way, in this chapter we privilege the conceptual exposition of our chosen framework over formal considerations. The formal exposition of our proposal will be expanded and discussed later in the next chapters.

Our holistic and integrative perspective rests on the fact that human disease is basically due to the disruption of the regulatory systemic processes that ensure the normal functionality of:

- cells,
- tissues,
- organs,
- and the human body as a whole,

as well as their feedback with environmental factors. Henceforth, we consider human health as the dynamical manifestation of a complex interplay between the large set of biological systems that constitute the body and the associated environment. Such manifestation occurs at the micro and the macro spatiotemporal scales as a consequence of alterations of systems-level developmental mechanisms.

© Springer International Publishing AG, part of Springer Nature 2018
M. E. Álvarez-Buylla Roces et al., *Modeling Methods for Medical Systems Biology*,
Advances in Experimental Medicine and Biology 1069,
https://doi.org/10.1007/978-3-319-89354-9_1

Biological systems, including developmental ones, are physical systems which exchange and process exchanges of matter, energy, and information. We introduce mathematical and computer-based formal modeling tools to describe and to understand the behavior of complex regulatory networks, which include both molecular and non-molecular components, and which underlie important aspects of the function of biological systems. Because of this, our conceptual approach belongs to the contemporary emergent *systems biology scientific domain of research*. And, since we are concerned by medical issues (i.e., we look for the improvement of human-health), our proposal belongs to what is currently known as the theoretical framework of *medical systems biology*. This domain of research constitutes an emergent field of biomedical scientific research. Because of its novelty, medical systems biology, sometimes also known as systems biomedicine (see for instance [281]), has not yet achieved a consensual definition. For our current purposes, we shall consider *medical systems biology* as the application of the conceptual framework of systems biology to the specific context of medicine.

We must point out that as a novel approach to deal with medical issues, the theoretical framework of medical systems biology can be applied any medical condition. In this volume, nonetheless, we are concerned only by a particular class of human diseases: the *chronic degenerative diseases*. The main reason that justifies our conscious choice is the systems-level relationship that intimately links chronic degenerative diseases with biological development. Our understanding of biological systems comes from both the theoretical and experimental study of the complex regulatory mechanisms that underlie developmental processes in multicellular organisms in time and space:

- Cell proliferation
- Cell differentiation
- Morphogenesis

Our aim is then to project our current knowledge on developmental systems-level mechanisms to the deep understanding of the emergence and progression of chronic degenerative diseases. With this, our main goal is to provide useful answers to questions such as:

- What leads to an altered cellular behavior or differentiation path?
- Why and how is tissue integrity lost?

We are motivated by using systems-biology approaches to eventually reinforce therapeutic strategies based on a preventive approach in medicine and public health. The rest of this chapter is organized as follows.

In Sect. 1.2 we overview the burden of chronic degenerative diseases (related to the complex interplay between synergistic risk factors and the human health state), and we justify why a systems-level approach is required to tackle them. In Sect. 1.3 we deal with the important issue concerning the synergistic interactions between risk factors that underlie physiological processes in health and disease, and how combined risk factors work through the disruption of regulation. Section 1.4

explores prevention as a therapeutic strategical choice, and we point out the limitations of classical pharmacological treatments.

We dedicate Sect. 1.5 to the key role that transcriptional core regulatory network modules play in cell fate processes. Having the emergence and progression of medical conditions in mind, we present there our main postulate:

> To understand how different components of regulatory networks (genetic and non-genetic, including random fluctuations, and intra- or extra-organismal environmental factors) shape healthy versus unhealthy conditions, both a biological developmental approach and a perspective based on complex systems must be taken into account.

This postulate is supported by the well-established connection between disrupted regulation and the consequent alteration of the genotype–phenotype mapping and altered cellular behaviors and identities. We also discuss the sources of cellular disruptions, and how the interplay between the gene regulatory networks (that underly cell–fate) with extra-cellular signals elicited by alterations in the cell's microenvironments, give rise to emergent transitions from normal to pathological phenotypes. In that section we also explain how our experience with models of plant systems guided us to uncover of some important generic processes that characterize biological development in multicellular organisms, and how the bottom-up systems-level understanding of these generic processes eases the uncovering of the dynamical phenomena underlying chronic degenerative diseases. We finish that section with an overview of the epigenetic landscape formalism, which structures in a formal way our medical systems biology bottom-up perspective by linking cell-level models with those that consider the temporal (and eventually spatial) patterns with which a particular cell type transits to different types or behaviors.

In Sect. 1.6 we justify why a systems-level state-based perspective (for the understanding of human health) allows the uncovering of the consequences of disrupted regulation in the onset of medical conditions. Moreover, we argue there how therapeutic strategies can be tackled in terms of goal-oriented regulated dynamics.

Finally, we conclude this chapter with a brief synthesis.

## 1.2 The Burden of Chronic Degenerative Diseases

According to the World Health Organization (**WHO**), chronic degenerative diseases, such as:

- cancer
- chronic respiratory diseases

- cardiovascular disease
- diabetes

affect up to 70% of the world populations and claimed around 50% of fatalities just in Latin America in 2012. These kinds of human diseases are characterized by:

1. The gradual aggravation of their symptoms, attributable to the continuous deterioration of affected tissues. This implies the loss of structural and systemic integrity.
2. The lack of effective treatments that can induce remission or achieve prevention. Once a chronic disease is present, it is a formidable challenge to cure it.
3. The negative side effects of current treatments. Pharmacological treatments intended to deal with chronic degenerative diseases are frequently designed to treat symptoms, not causes, and may even be detrimental for the organism.

These three main characteristics of chronic degenerative diseases are responsible for a substantial part of the global social and economic burden that these diseases currently represent. They are manifested through the increase of the treatments costs and efforts, as well as human suffering. Consequently, considerable scientific and technological research efforts are being currently devoted to try to shed light on the complex mechanisms underlying:

- onset,
- progression,
- treatment,
- and prevention

of these diseases. In this sense, research has mainly focused on elucidating how a given specific chronic degenerative disease arises from individual disturbances on the intricate cellular and biochemical machinery regulating normal physiological functions. These disturbances include disruptions of components of the involved regulatory networks as well as alterations of the interactions between components. From such research it has become clear that most complex chronic degenerative diseases can be caused by *several different* factors such as:

- Chronic inflammation [308].
- Alterations of microbiome [54].
- Modulation of lifestyle [321].
- Genetic mutations or polymorphisms [286].
- Pollutants and other nocive environmental factors (e.g., ultraviolet irradiation) [170].
- Toxic substances (e.g., organophosphorus pesticides in fresh vegetables, see for instance [509]).

Furthermore, it has been stated that disease can be triggered by different combinations of risk factors (synergism), with different severities, and with a differential role depending on the stage of the concerned disease [4, 135, 483]. Thus, it is difficult to pinpoint individual pathogenic triggers that are responsible for the onset and progression of complex chronic degenerative diseases. Instead, it is becoming clear that chronic degenerative diseases emerge from complex interplays between different possible genetic, non-genetic intracellular and extracellular but intra-organismal, as well as environmental (extra-organismal) risk factors [57, 263]. Synergism must be taken very seriously. Even if particular risk factors are accessible to intervention, it does not necessarily mean that the provoked detrimental consequences can be easily erased or reverted.

We can visualize human health and disease as a collection of specific well-differentiated *states* that characterizes the concerned physiological state disease or medical condition. The concept of state, associated to a given system, is the perceived condition of that system at a specific given time. Thus, a disease can be described enunciating the corresponding states (i.e., the time-dependent circumstances that define the disease under analysis). Thinking in terms of states, describing specific time-dependent phenomena in terms of state trajectories is then an intuitive cognitive task. Therefore, a state trajectory is a time-indexed collection of health conditions. In the same way that we say that a physical particle motions in a physical space (where the trajectory of the particle, resulting from the action of forces, can be described by a time-dependent collection of coordinates including positions and velocities with respect to a given inertial frame), we can say that the health of a given person motions in a health state-space. As we shall see in a later chapter, the conceptual framework proposed here can be enunciated in a formal quantitative manner. At this stage of the volume we shall avoid formalization, however.

Before presenting the discussion turning around how the concept of state can be applied in order to organize a systems-level treatment of human-health phenomena, we shall be more specific in what follows on the interplay between risk factors and human disease.

## 1.3  Risk Factors of Human-Disease

### *Synergistic Interactions Between Risk Factors*

Synergistic interactions between risk factors result from the high spatiotemporal interdependence between the:

- biochemical,
- cellular and
- tissue-level

reactions that underlie the physiological processes in health and disease (see Fig. 1.1). In order to clarify this statement, let us consider, in what follows, two well-known illustrative chronic degenerative disease examples, *colorectal cancer* and *diabetes*:

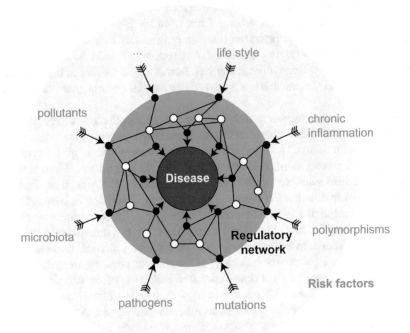

**Fig. 1.1** Risk factors. Chronic degenerative diseases emerge from a dynamical complex interplay between genetic, physicochemical, stochastic, environmental, and microenvironmental risk factors. This interplay blurs the distinction between cause and effect because of the presence of feedback-based dynamics

**Colorectal Cancer and atopic diseases:** It has been hypothesized that the onset of colorectal cancer can be associated to the contact of altered intestinal microbiota with epithelial cells bearing a mutation in the mucus-encoding *Muc*2 gene [409]. Thus, at least both the genetic and the environmental factor must be present to initiate the uncontrolled proliferation of epithelial cells that characterizes the early phases of this cancer. The presence of either the mutation or the environmental factor is not enough to trigger this disease phenotype state, because the intercellular signaling networks controlling the progression of the cell-cycle have intricate and robust regulatory structures that can buffer out single but not co-occurring perturbations [426]. Such an increase in the susceptibility to environmental risk factors, in the presence of genetic alterations, also seems to underlie the onset and progression of allergic diseases, such as atopic dermatitis, asthma, and inflammatory bowel disease (see [122, 366, 411]). It is due to:

- The dynamic interplays that exist between the epithelial barrier function (epidermis, airway epithelium and intestinal epithelium, respectively).

- The environmental stress factors to which these tissues are exposed (pathogens, pollutants, or food components).
  and:
- The resulting immune responses that are naturally triggered by the excessive disruption of epithelial tissues by environmental aggressors (see [132, 299, 458]).

**Diabetes:**   The elevated levels of glucose are associated to the disruption of inflammatory processes. The increase of the rate of inflammation desensitize the insulin response to glucose in pancreatic cells, further increasing the glucose levels [78] (this phenomenon is discussed later in this volume). This complex interplay between glucose levels and inflammatory processes shapes a negative feedback loop. The detrimental consequences can be further aggravated by a myriad of different genetic conditions [154, 178].

These couple of examples illustrate that a given chronic degenerative disease in fact emerges from altered complex networks including both genetic and non-genetic components.

*Remark 1.1 (Cause and Effect Distinction Blurred by the Existence of Feedback-Based Dynamics)* As pointed out by the previous illustrative examples, the existence of feedback-based interdependencies blurs the distinction between causes and effects. Bottom-up approaches of medical systems biology addresses the elucidation of the dynamical significance of feedback-based interactions between genetic and non-genetic components, avoiding simplistic (and dangerous) quests for a single cause of a given complex disease.

To understand how the different combinations of risk factors contribute to the development of cellular disease-characteristic phenotypes, it is therefore crucial to understand how they are functionally connected to the regulatory networks controlling tissue function. Let us now proceed to tackle this issue.

## *Interconnectivity of Risk Factors*

Many of the current clinical challenges in the field of chronic degenerative diseases are a direct consequence of the strong functional interconnectivity between risk factors, requiring integrative, quantitative, and dynamic approaches, in order to:

- Elucidate how disturbances elicited by risk factors combinations propagate across regulatory networks underlying cell behavior and tissue integrity.
- How this propagation of perturbations leads to a gradual deterioration of the health-state (i.e., disease condition).
- How early and pre-clinical phases can be identified from network properties.

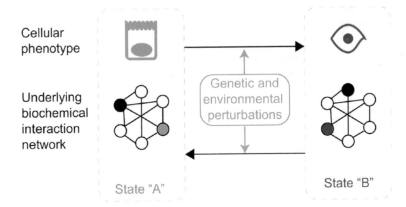

**Fig. 1.2** Cellular plasticity. Different phenotypes of individual cells correspond to different states of activation of complex biochemical interaction networks. Genetic and (micro)environmental factors can drive changes in the phenotype of cells by interfering with these complex biochemical interaction networks

- How a systems-level and quantitative framework can be used to design optimal treatment intervention strategies for halting or reverting the progression of a particular disease.

The first consequence of the strong functional interconnectivity between risk factors is the synergism between them, as was pointed out previously. This can be attributed to the specific properties of biological regulatory networks that tend to be able to compensate single perturbations [10, 72, 161, 226, 227] but are often vulnerable to combinatorial attacks [121, 363, 425]. Further, due to the connectivity among the regulatory network components, several different combinations of risk factors can converge to a phenotypic transition from healthy to disease states with conserved symptoms [122, 316]. We hypothesize that such convergent or generic patterns emerge from the modulation of *core regulatory networks* that underlie the cellular transitions in both normal and ill conditions in response to various perturbations, including genetic alterations [112, 217, 370, 371] (Fig. 1.2). This issue will be discussed in detail throughout this volume.

*Remark 1.2 (Combined Risk Factors Work Through the Disruption of Regulation)* A systems biology approach, which explicitly considers the regulatory network mediating the interplay between the many potential targets for genetic and environmental risk factors, is essential to systematically characterize the role of different combinations of risk factors on the onset and progression of disease. Robustness against single perturbations is a common property of regulatory networks, but fragility of regulatory networks is also common when simultaneous risk factors are present or specific components with particular positions within such networks are altered. The balance between fragility and robustness is maintained in health but lost in disease conditions and such systems-level behaviors are an important focus of systems biology research.

The second consequence of the strong interactions among the different genetic players (physicochemical components underlying tissue function) is the gradual deterioration of the disease phenotype. It occurs through the regulatory logic that mediates the interplays between the phenotypes of cells in the tissue and the microenvironment they are embedded in. While microenvironmental conditions can modulate cellular phenotypes [26], the distributions of cellular phenotypes within a tissue in turn alter the microenvironment. Hence, while this interplay between cellular phenotypes and microenvironments ensures a coherent adaptation of both microenvironments and cellular phenotypes, alterations in this feedback structure, for example by genetic or environmental factors, can trigger a rapid progression of the disease with fatal consequences. The following examples illustrate this complex process:

**Fibrotic Diseases:**   During the wound healing process, epithelial cells trans-differentiate to a mesenchymal phenotype in response to increased levels of the TGF$\beta$ cytokine, which is released as part of the inflammatory response to tissue damage [230]. The accumulating pool of mesenchymal cells secrete additional TGF$\beta$, potentiating the induction of the epithelial-to-mesenchymal transition [348, 497]. Indeed, alterations in this positive feedback affected by mutations, polymorphisms, tissue-damaging drugs [174, 508], and many other factors, can lead to fibrotic diseases including:

- Cirrhosis
- Pulmonary fibrosis carcinomas
- [394].

**Induction of Blood Vessel Formation by Growing Tumors:**   Hypoxic   stress can be generated by the uncontrolled growth of tumor cells that move away from the nearest blood vessel as they continue dividing. This low oxygen condition triggers an adaptation of these cells, from a high-proliferative state to a more aggressive cell phenotype that is motile and resistant to chemotherapy [94]. Simultaneously, hypoxic conditions induce the production of new blood vessels via the release of the growth factor VEGF [261], and the eventual restoration of normoxic conditions [27]. As a consequence, although tumor cells can re-differentiate back into a less malignant state, also their high proliferative potential is restored, and they can invade new organs by the newly formed blood vessels [94, 455].

**Progression of Atopic Dermatitis:**   This disease can occur as a consequence of repeated rises in the cytokine levels in response to pathogens that have invaded the viable layers of the epidermis. Increased cytokine levels weaken the epithelial tissue, leading to a further colonization of viable epidermal layers by infiltrating pathogens and with that, a gradual deterioration of epidermal function [122, 123] (this example will be extensively discussed ahead).

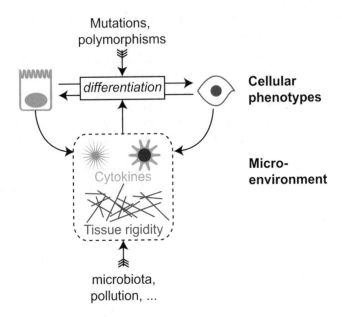

**Fig. 1.3** Microenvironmental factors. Complex interplays between cellular phenotypes microenvironmental factors can lead to the gradual aggravation of the disease

*Remark 1.3 (The Key Role of Microenvironment)* As illustrated by the previous examples, a systems biology framework to understand the complex interplay between cellular phenotypes and their microenvironment is important to elucidate the mechanisms for onset and progression of complex diseases. Furthermore, the transition between the different disease stages can be viewed as an aberrant developmental process, in which the progression from a healthy, pre-clinical state to a severe disease phenotype emerges from the complex interactions between the cellular phenotype and the associated microenvironment (see Fig. 1.3). Systems biology approaches allow then the systematic analysis of the environment-phenotype interactions, as well as the effects of genetic and environmental perturbations on this regulatory structure, from an integrative, mechanistic, and dynamical perspective.

Since the interplay between the environment and biological regulatory networks underlie functional plasticity, medical therapeutic strategies can be developed in order to modulate the interplay between environment and regulation. Therefore, prevention therapeutic strategies intended to modulate the interplay between the environment and the (disrupted) regulatory processes are a sensible choice. In what follows we shall discuss this issue.

## 1.4   Prevention as a Therapeutic Strategy

### *Early Warning Signals*

In terms of finding effective prevention strategies, it is also natural to view as a dynamical process the progression of chronic degenerative diseases. Finding effective prevention strategies then corresponds to halting the natural progression from a healthy state to a *susceptible but pre-clinical stage*, and from this to a disease phenotype (see Fig. 1.4). In this sense, the first challenge is to be able to identify and characterize the high-risk patients, who do not present symptoms but would strongly benefit from early intervention strategies that can effectively halt the incipient onset and progression of disease. In other words, to prevent the progression of degenerative chronic disease it is important to find early markers or signs that distinguish cohorts of vulnerable patients while they are still at a healthy or a pre-clinical stage. Since such patients are still asymptomatic, it has been hard for medical practitioners to find early bio-markers in the form of individual molecular species that effectively predict an incipient onset of a severe disease. Recent work in the field of systems biology has shown that such *early warning signals* in general do not depend on individual genes or proteins, but can be determined from:

1. The global network properties that can be obtained from high-dimensional profiles of gene expression [84].
2. Subtle quantitative changes in the dynamical behavior of the underlying regulatory networks [149, 449, 450].
3. Increased variance and decreased recovery rate of some regulatory network components [122].

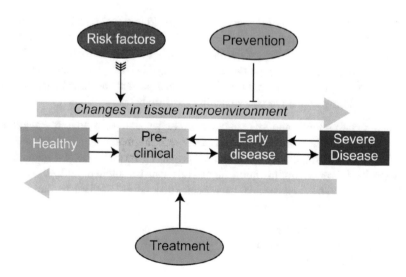

**Fig. 1.4** Modulation of pathology. Pathology as dynamical process that can be modulated by treatment and prevention strategies

Finding early bio-markers for the identification of vulnerable patient cohorts can thus be seen as an essential challenge in understanding and preventing chronic degenerative diseases, and which can fortunately be tackled from the networks-based perspective of systems biology.

## Unwanted Side Effects of Pharmacological Treatments

A current clinical problem in treating chronic degenerative diseases is the unwanted side effects of pharmacological treatments, which can occurs as a consequence of the lack of target specificity [327], leading to the interference of the drug with normal physiological processes [407]. This effect can be further aggravated by the desensitization of the affected tissue exposed to the drug [199, 377], through which increasing doses or variants of the drugs are required to relieve the symptoms as the disease progresses.

Since drugs act on organisms through complex regulatory networks, pharmacological treatments are not free from risk. This clearly emphasizes the advantage of preventive therapeutic strategies. Nonetheless, pharmacological treatments are frequently unavoidable. Taking into account the risks, optimal treatment regimens of minimal durations and minimal doses that minimize the unwanted side effects but effectively induce remission are required. Designing such treatments, however, is a challenging task that requires the consideration of the many regulatory interactions between the molecular components underlying the healthy and disease phenotype, many potential drugs combinations with different strengths and durations, and the wide spectrum of possible triggers and states of the concerned disease (recall Fig. 1.1).

*Remark 1.4 (Risky Pharmacological Treatments)* Pharmacological treatments can also constitute by themselves a risk factor that can be taken into account. For instance, a 52% increased risk of mortality associated with tiotropium mist inhaler in patients with obstructive pulmonary diseases [13] has been reported. Also, treatments with pioglitazone and rosiglitazone can produce adverse cardiovascular events when being applied to patients with type 2 diabetes.

Thus, systems-level and dynamical approaches can help to circumvent the conundrum resulting from the interplay between pharmacological treatments and complex networks regulatory dynamics by providing mechanistic and quantitative frameworks to which optimization algorithms can be applied. Indeed, recently, such systems biology approaches have been successfully applied to find optimal treatment strategies that stably revert:

- Prostate cancer [198–200, 407]
- Hepatocellular carcinoma [425]
- Atopic dermatitis [88]

These recent results exemplify the therapeutic potentialities of medical systems biology approaches.

We can at this level examine how the underlying genetic and non-genetic components of the human system map unto phenotypes that correspond to healthy, preclinical, and diseased conditions or states. This will lead us in what follows to the consideration of a particular type of transcriptional regulatory network: *core regulatory modules*.

## 1.5 Multistable Transcriptional Core Regulatory Network Modules

Although medical conditions as the ones considered up to now are generally characterized at the phenotypical level, a systems-level biological approach to understand their emergence and progression necessarily implies the study of how the underlying genetic and non-genetic components of the human system map unto phenotypes that correspond to healthy, preclinical, and diseased conditions or states. Gene regulatory networks composed mainly of transcriptional factors have been shown to be particularly important in development and hence in mapping mechanisms from genotypes to phenotypes. Such gene regulatory networks can attain various steady-states that relate to different phenotypes and are hence multistable dynamical systems. These may attain different states as a result of alterations of their components, but also in response to environmental cues [5, 271].

Here, we refer to gene regulatory network modules that underlie cell transitions that are key for particular developmental processes under health and disease as *core regulatory modules*.

### *Genotype-to-Phenotype Mapping and the Emergence and Progression of Medical Conditions*

In order to understand how the genetic and non-genetic components of organisms are implied in the emergence and progression of disease, it is important to understand how different genetic or other types of molecular components, (e.g., transcripts, regulatory sequences, proteins, and metabolites), collectively function at different levels of organization, such as:

- Coupled networks
- Cells
- Cellular microenvironment
- Tissues
- Organs

In addition to the molecular, genetic, and non-genetic intra-organismal conditions, recent studies are also documenting the key dynamical and fundamental role of stochastic fluctuations (see for instance [215, 217, 304, 316, 372]), as well as extra-organismal environmental factors that can be associated to the lifestyle (see for instance [291] and [487]) and also to the exposure to geno-toxic factors (radiation, toxic chemicals, etc.; see for instance [445]) that may or may not be related to a person's lifestyle (see for instance [119]).

A longstanding tradition in the biological developmental field of research has focused on studying the concerted action of multiple genetic and non-genetic components (i.e., physicochemical constraints), together with environmental factors during cell differentiation and morphogenesis (see [240, 446, 457, 473], among others). Such general perspective is applied here to specific biomedical cases using a bottom-up approach to systems biology.

*Remark 1.5 (Our Main Postulate)* The main postulate of this volume is that in order to understand the role of different types of components (genetic and non-genetic, including random fluctuations, and intra- or extra-organismal) in underlying healthy versus unhealthy medical conditions, both a biological developmental approach and a perspective based on complex systems must be considered. As health conditions emerge from the feedback-based interplay between intra-organismal complex networks or systems and extra-organismal or environmental factors, the role of genetic components (and/or environmental conditions) can only be understood on the basis of their collective action in the context of highly nonlinear and multistable dynamical regulatory networks (here multistability refers to the property of a network having multiple circumstances of equilibrium). Moreover, the fundamental role of noise or stochastic fluctuations can only be understood in the context of such networks and the involved so-called stochastic resonance [307] that emerges from the interaction among deterministic and non-deterministic dynamics of the system. Additionally, feedback-based dynamics between developmental and metabolic processes also shape genotype–phenotype interactions (see Fig. 1.5).

A complex systems approach requires mathematical and computational models that enable analyses of the collective action of the implied components dynamically, and also of how such collective action yields emergent systems-level structures and behaviors. The sets of interactive genes, proteins, other biomolecular players, and the cellular phenotypes and tissue-level properties they control, constitute complex dynamical regulatory networks. Depending on the architecture (and nature) of involved interactions, these networks can be plastic, resilient, and/or robust, which are clinically important properties of the system that can be studied only from an integrative and dynamical perspective.

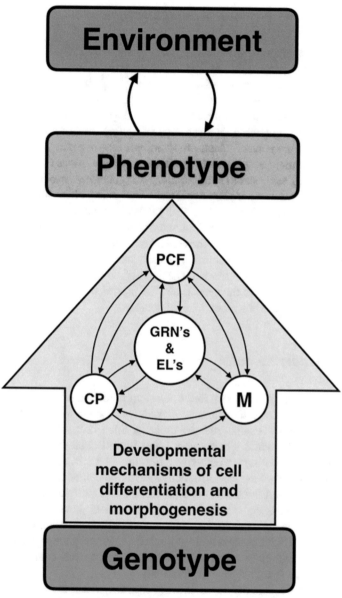

**Fig. 1.5** Genotype-phenotype mapping. This schematic representation shows how the genotype is mapped into the phenotype through developmental mechanisms, under the constraints imposed by environmental feedback-based interactions. Therefore, cell differentiation and morphogenesis result from the interplay between: developmental gene regulatory networks and the associated epigenetic landscapes, cell proliferation dynamics, physico-chemical fields, and the dynamics of metabolism. Phenotypic variation is influenced by environmental factors (e.g., temperature, humidity, and stress)

## *Looking for the Causes of Cellular Disruptions*

In contrast to the genomic top-down approaches that exhaustively search for all the components (transcripts, proteins, metabolites, etc.) and rely on descriptive, and mostly static, as well as associative approaches (see for instance [59, 138, 275, 276, 391]; and for a general discussion see also [150, 246, 452]), we follow in this volume a systems-level bottom-up approach. Such perspective implies integrating from well-curated functional data for a set of components and their interactions. These comprise core gene regulatory modules with the necessary and sufficient set of restrictions required to recover observed stable configurations for the components included. Each of such stable configurations are generally associated to different cellular state or other level of phenotypic state. Such relatively small gene regulatory network modules are analyzed dynamically to understand the systems-level mechanisms involved in both normal and altered states at the cellular, tissue, organ, and organismic levels (see Fig. 1.6). In the context of medical systems biology, this class of gene regulatory modules, comprising mostly transcriptional regulators, underlie specific developmental processes that when altered, are implied in the emergence of cellular states associated to the genesis and the progression of human diseases. Let us now tackle this issue in a more detailed manner.

## *Core Regulatory Modules*

We consider that the critical step in a bottom-up modeling effort implies deciding which components and interactions to consider and when to propose a new core regulatory module to be analyzed dynamically. Such task is generally easier for well-characterized processes or cases of phenotypical alterations at the molecular level. Bottom-up modeling is a recursive procedure, in which simulated stable states and observed profiles of expression or activation of the components of consideration are iteratively compared. Once a set of necessary and sufficient set of restrictions have been integrated, we say that a robust dynamical gene regulatory network module has been uncovered. During such recursive procedure, experimental holes are uncovered and in the context of the proposed dynamical gene regulatory network module, novel predictions of necessary interactions or even components can be proposed and later searched for experimentally.

Transcriptional regulatory modules are particularly relevant for integrating various signals and also concerting cellular decisions or fate decisions. In our previous studies on plant development, we have shown that such transcriptional regulatory cores or modules (see Fig. 1.6) are key shaping patterns that underlie:

- Cell–fate decisions.
- Tissue patterning.
- Morphogenesis.

(see for instance [28, 29, 141, 318–320]; for a review on this subject see [19]).

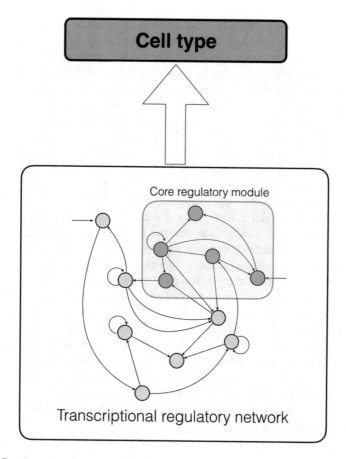

**Fig. 1.6** Core gene regulatory modules. Empirical evidence shows that transcriptional gene regulatory networks are modularly structured. In networks regulating biological development we are able to identify robust modules, which we call core regulatory modules (e.g., early mesoderm development in Drosophila melanogaster [398]; mesoendoderm specification in mouse [48]; neural crest development in vertebrates [419]; flower development [141]; differentiation of retinal pigmented epithelium [384]). Feedback-based interactions play a key role in core gene regulatory modules (e.g., feedback circuits regulate the number and size of attractors, see for instance [31] and the references therein)

Hence, such multistable regulatory cores are also very important in understanding why and when a genetic or non-genetic alteration yields a phenotypic modification. A guide to finding relevant regulatory core modules comes from the existence of conserved or generic patterns and behaviors. This suggests the existence of underlying robust systems, that may be the regulatory modules. Uncovering such modules from well-curated functional data is an important first step and contribution to understanding the systems-level mechanisms that underlie both the conserved and variable phenotypes under analysis, or the *healthy and unhealthy medical conditions*.

In this volume we propose to follow a similar approach to that used in plant systems. We aim at uncovering relevant transcriptional regulatory modules or cores that may underlie documented conserved or convergent patterns of cellular or tissue behavior under normal and pathological conditions that suggest underlying robust systems-level dynamical mechanisms. We propose that complex human unhealthy conditions may emerge from alterations (genetic and non-genetic) of the same multistable regulatory networks operating during normal development.

The regulatory cores of interest may be integrated from data under normal conditions in humans and also from animal models used in experimental molecular genetic studies; under the assumption (based on experimental data) that such transcriptional modules are generally conserved among related species [244, 278, 392]. Such regulatory cores are important building blocks in the systems-level mechanisms that underlie the emergence of aberrant cellular phenotypes.

Since biological systems are non-isolated from their surroundings, regulation is necessary to modulate the interaction between the concerned cell systems and the environment. Henceforth, as an evolutionary result, transcriptional regulatory networks fulfill complementary tasks: regulation of the cell functional identity overlapped with the regulation of the cell response to exogenous stimuli. These modulatory tasks then condition the structural and functional properties of cell regulatory networks, fixing specific functions for specific nodes (e.g., interconnectivity of the network with the environment). Bottom-up medical systems biology approaches lead to the development of systems dynamics methods that identify the specific role of particular nodes. This is particularly appealing when looking for the dynamical consequences of the interactions between transcriptional regulatory networks and signal transduction processes (that mediate the interplay between the regulatory networks and the environment).

*Remark 1.6 (Cell-to-Cell Interactions)* We must remember that in multicellular organisms cells interact to fulfill collective tasks that drive tissue-level function. This means that cell-to-cell interactions (mediated by extracellular signaling processes) are also concerned by intracellular transcriptional regulation (and sometimes also at the intercellular level (see for instance [159, 378]). The cell-to-cell communication machinery allows the coordination of cellular communities, and the disruption of this machinery may be also involved in the emergence and progression of chronic degenerative diseases. The understanding of the phenotypic consequences of disruptions of the complex couplings between intracellular regulatory networks and extracellular structures and processes is an important target for medical systems biology research.

In what follows we discuss how extracellular signals and microenvironments interact with (intracellular) regulatory networks, and how the disruption of the involved coupling processes give rise to pathological dynamics.

## *Extracellular Signals and Microenvironments*

The feedback-based interactions with extracellular signals and microenvironments, as well as the coupling with other cells within tissues, are fundamental (see for instance [232] and [123]) to understand biological development and the corresponding dynamics of chronic degenerative diseases.

While the state of the intracellular networks that couple transcriptional and other types of regulation, including epigenetic mechanisms, with signal transduction factors respond to microenvironmental factors, these are also affected by the cellular states or types (see the example concerning atopic dermatitis in the corresponding chapter). Hence, we aim at integrating such models of intracellular regulatory network that are multistable and help understand the dynamics of:

- cell differentiation;
- cell reprogramming;
- or cell dedifferentiation,

implied in pathological processes, with mechanistic models that incorporate physicochemical fields (see Fig. 1.7). This implies developing models that consider the feedback with other cells, but also with:

- Extracellular signals
- Chemical and physical fields
- Extra-organismal environmental factors

As stated previously, we postulate that it is in fact from the feedback of the intracellular systems and the environment that the different normal and pathological phenotypes and their transitions emerge. Feedback-based interactions are fundamental for the robustness and plasticity dynamics that rule biological regulatory processes and thus both healthy and diseased phenotypes.

Overall, we propose in this volume mathematical and computational approaches that are useful to understand the emergence of observed conserved patterns at different levels of organization during normal and pathological conditions. The restrictions that underlie conserved, repetitive, or generic patterns emanate from mutual interactions and feedback-based mechanisms at the genetic and non-genetic levels, as well as across temporal and spatial scales as is explained ahead in this book. These multi-level and multi-scale processes comprise the biological mechanisms that underlie cell differentiation and cell reprogramming, which in turn underlie tissue patterning and morphogenesis under both normal and pathological conditions [19].

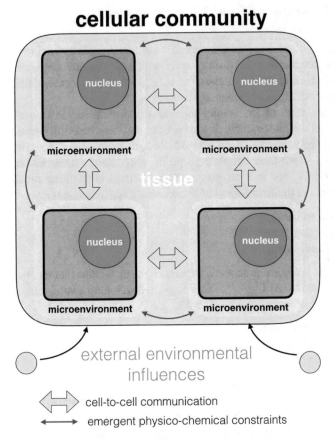

**Fig. 1.7** Cell-community. Transcriptional regulation coordinates the biomolecular interactions that give rise and defines cell-identity. In multicellular organisms cell-identity strongly depends on the cell-community to which it belongs. Cellular couplings, as well as cell-to-cell communication, are conditioned by transcriptional regulation. Moreover, the interaction of the cells integrated into a given cell community give rise to the emergence of physicochemical fields or constraints, that also feedback to the intracellular networks dynamics and modulate the influence of external environmental factors. The disruption of this complex feedback-based network can challenge cellular identities, potentiating then the emergence of disease as previously described

## *Limits of Top-Down and Gene-Centric Approaches*

We aim at an approach that complements top-down approaches that are becoming particularly informative as single-cell and laser microdissection methods are being used to characterize cell profiles and phenotypes under particular health conditions at the level of transcripts, miRNAs, proteins, metabolites, and so on (see for instance [53, 302, 311, 440] and the references therein). Such top-down approaches are also aimed, beyond description of patterns encountered, to infer systems-level mechanisms involved. Using bottom-up approaches it may be possible to identify gene regulatory network modules that dynamically underlie the configuration changes uncovered in cells at different stages of a disease progression. It is the collective action of the complete network's interacting molecules in conjunction with non-genetic components, such as physical and chemical fields and the associated environment, that underlie normal and altered cell differentiation, pattern formation, and morphogenesis. Nonetheless, the bottom-up approaches may aid at uncovering some of the key modules of such interactomes.

In conclusion, the "top-down" and the "bottom-up" approaches complement each other and increase the explanatory and predictive power of systems biology (see [67]). A key aspect of the bottom-up systemic and mechanistic approach is that it considers explicit and experimentally grounded nonlinear dynamical mapping models at different levels of organization (e.g., gene interactions and circuits, networks, cells, tissues, organs, etc.) incorporating both molecular-genetic (e.g., genes, proteins, miRNAs, etc.) and non-genetic (e.g., mechanical and elastic forces and fields, chemical concentrations and gradients) components to understand the genotype–phenotype mapping and to uncover systems-level behaviors and traits. Such multistable dynamical models view cellular behavior and cell differentiation as the inevitable manifestation of the intrinsic nonlinear and stochastic nature of underlying networks of interacting components. Hence, we postulate that such an endeavor implies uncovering the biological mechanisms of cell differentiation and morphogenesis under disease and healthy states. In summary, in our view, bottom-up systems biology is fundamental for a complex systems and dynamical mechanistic approach to the understanding of health issues.

## *Nonlinearity and Stochasticity*

The importance of the nonlinear and stochastic nature of biological systems has been emphasized before (for example [240, 312, 401, 457]). More recently, dynamical models have been validated for particular biological systems by integrating mechanistic data for the system's components, their interactions, and their concerted action, as well as the emergent structural and dynamical consequences of such complex systems (see for instance [42, 69, 318]). In this volume we illustrate the constructive role resulting from the inevitable interplay between feedback-based interactions among genetic and non-genetic components and stochasticity in real biomedical conditions. We must point out that the latter view has been previously put forward in different terms by pioneer theoretical biologists and their application to development and evolution (see [172, 241, 422, 473]).

*Remark 1.7 (Avoiding Reductionism)* Traditional cause–effect, reductionist or linear explanations in biology and biomedicine do not consider explicitly the conditional role of each part on others, or the direct and indirect interactions among components that altogether comprise the system (see, for example, discussions in

[16, 211, 232]). We consider that a systems-level approach that explicitly considers interactions and the nonlinear stochastic nature of underlying processes is required to understand the emergence of the observed biological patterns. Hence, for the purpose of this volume, by reductionist paradigm we refer to an explanatory view in which the effect of interactions among components of interest (e.g., molecular species) is (in)advertently ignored.

### → Plant Systems as Developmental Models for a Bottom-Up Systems Biology Approach

It is important to explain here that part of our bottom-up systems biology approach has been built on the study of plant biological development. Plants are very useful model systems for experimentally grounding and validating theoretical models. In fact, we have tested our proposal for over two decades of research on complex network models applied to plant systems, mainly *Arabidopsis thaliana* (see Fig. 1.8). See the review [19], and the references therein, for a detailed discussion on this issue, and also to explore some of our examples concerning plant developmental biology, for example:

- Flower development
- Spatial cell patterns in *Arabidopsis thaliana* root
- Cell patterns emergence from coupled chemical and physical fields with cell proliferation dynamics.

As we shall see in the next chapters, the bottom-up systems biology approach for the understanding of biological developmental phenomena can be useful to understand the systems-level mechanisms implied in chronic degenerative diseases. As evidenced by theoretical and experimental research, some generic systems-level mechanisms underlie fundamental aspects of the complex developmental processes of multicellular organisms (e.g., cell differentiation and cell morphogenesis). These mechanisms are self-organizing and have been attained in convergent evolution in different evolutionary lineages (see for instance [17, 181, 495]). This explains why we can uncover human developmental processes under both health and disease with similar approaches to those used to uncover systems-level mechanisms of plant and animal development.

It is now time to present a useful tool for the understanding of the emergence of temporal and spatial morphogenetic patterns as a result of the developmental regulatory networks: the *epigenetic landscape formalism*.

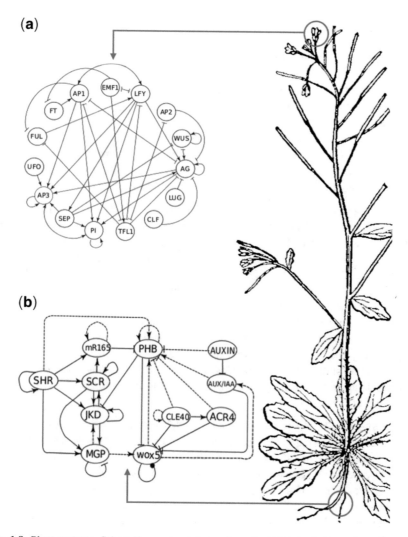

**Fig. 1.8** Plant systems. Schematic representations of an *Arabidopsis thaliana* plant (modified from [460]). Two developmental processes of this model organism that have been studied using multistable gene regulatory networks are indicated: flower morphogenesis and niche organization of the root stem cell. For each example, the corresponding gene regulatory networks were characterized via available empirical evidence. (**a**) Depicts the gene regulatory network underlying flower development (see [16, 141]). (**b**) Shows the gene regulatory network underlying root stem cell niche organization (see [32])

## *Epigenetic Landscape*

### → **Attractors and Cellular Identities**

The paradox that contrasting stable cellular phenotypes associated to different fates or types are manifested despite an underlying invariant genomic sequence clearly indicates that cellular differentiation is a consequence of *epigenetic regulatory mechanisms*. By this, we mean that the consolidation of a given cell phenotype (i.e., the cell-identity at a given time), is *not only* dependent on the genome sequence of the cell. The identity of cells with different fates result from complex systems-level self-organizing mechanisms, including regulatory networks.

*Remark 1.8 (Stable Cellular Phenotypes as Dynamical Attractors)* If cells are understood as dynamical systems that can be described in terms of a chosen state variable, a particular cellular phenotype can be then seen as an *attractor*. As its name suggests, an attractor corresponds to a robust circumstance of phenotypic equilibrium. To say that a given system's state is an attractor, means that the concerned system remains in that perceived condition at a specific time *even in the presence of disturbances*, i.e., the condition is attractive (see Fig. 1.9). This stable condition is guaranteed by the underlying regulatory network and its restrictions or interactions. Under some circumstances the stable condition can be altered, and consequently the system will move to a different state. An invariant regulatory network may attain several attractors and it hence constitutes *per definition* a multistable system. Thus, a given system is said to be multistable if it possesses more than one attractor.

The dynamical models proposed later in this volume constitute a mechanistic explanation for the emergent stable gene configurations or gene expression profiles that are associated to different cell types. As we shall see, the proposed gene regulatory networks are in fact multistable, with each stable state or attractor interpreted as a particular configuration of gene expression associated to each cell type (see for instance [240, 264, 318]) or eventually health state. It must be pointed out that multistable systems are necessarily nonlinear (see for instance [138]).

In historical terms, Stuart Kauffman first used randomly generated Boolean networks (in which each component may attain only two values: 0 and 1) to explore various dynamical network cellular behaviors with contrasting structures randomly generated in silico. However, such network structures were not similar to those found in actual living cells. Indeed, later, the first network models grounded on experimental data showed that the structure of biological networks are different to that of random ones (see [9, 318, 472]; for some reviews on this subject see: [16, 109]). In the context of such network models, the "functional role" of each gene can only be defined *and understood* in the context of its interactions and beyond: in terms of the system where it is immersed. Moreover, each gene in the network is simultaneously subject to this regulatory process; consequently, global, non-intuitive, and structured behavior emerges in a self-organized manner.

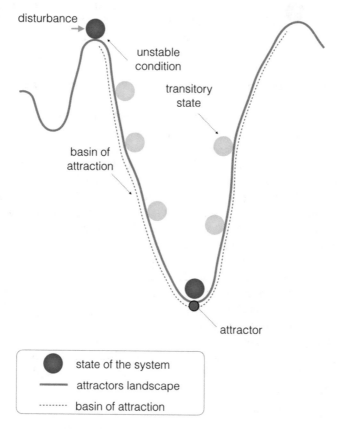

**Fig. 1.9** Attractor notion. A dynamical system may be in stable circumstance or in an unstable one. A stable circumstance is known as an *attractor*. If the system is in such a condition, it will remain there even in the presence of (transient) disturbances (up to some limit). In an unstable condition even low-intensity disturbances will more the system away from its attractor. Between two conditions of equilibrium, the system displays transitory states. For a given attractor, the system will be associated to a basin of attraction. For a given dynamical system, the union of the attractors with their respective basins of attraction is defined as the attractors landscape of the system

In summary, such complex regulatory networks converge to a discrete number of genetic configurations that are consistent with the restrictions imposed by the underlying gene regulatory network. Once such simulated configurations or stable states coincide with those observed experimentally, a regulatory module can be postulated (some reviews and examples are given in [16, 17, 213]). Qualitatively, cell fates or phenotypes can be associated to the in vivo manifestation of attractor configurations of the global gene regulatory network and in fact, this has been experimentally validated using human cells (see [214]). Such networks are experimentally validated with robustness analyses that test to what extent the network's attractors resist alterations of the interactions proposed. Also, the recovered attractors are compared to those of mutant lines, once the corresponding component is either fixed to 0 or 1 to simulate loss and gain of function mutations, respectively (see ahead).

### → Epigenetic Landscape and Morphogenesis: Populations of Cells

Gene regulatory networks also restrict the transitions among attractors. Such restrictions depend on a nonlinear and multidimensional (with as many dimensions as components, i.e., nodes, are in the network under consideration) quasi potential function that restricts the transitions among attractors. Such quasi potential corresponds to what Waddington called the epigenetic landscape [214, 473]. More detail on the formalities implied in the derivation, quantification, and analyses of the epigenetic landscape will be provided in the following chapter. For now, we want to state that the epigenetic landscape restricts the temporal and spatial transitions among attractors. The analyses of the epigenetic landscape that emerges from a particular regulatory network can be derived and analyzed in the context of both deterministic and stochastic models. The first analysis of the epigenetic landscape for a biological network grounded on experimental data was done in the context of stochastic explorations of the transitions from one attractor to another one (see [18, 468] and the references therein).

Stochasticity is another important property of biomolecular systems (see for instance [442]). Interestingly, the apparently repetitive and deterministic patterns of development seem to emerge from the feedback between complex intracellular gene regulatory networks with:

- Deterministic dynamics
- Nonlinear interactions
- Under the influence of stochastic fluctuations

Stochasticity results from either intrinsic or extrinsic fluctuations in regulatory interactions, or sampling errors resulting from limited numbers of molecules involved in such interactions (see [21, 63]). Stochastic resonance that results from such stochastic nonlinear systems seems to underlie the emergence of important biological patterns [18, 353, 511].

A stochastic exploration of the epigenetic landscape was first achieved for the regulatory network that underlies floral organ specification during early flower development [18]. The temporal pattern with which floral organs emerge during flower development is widely conserved among close to a quarter of a million of flowering plants. Figure 1.10 shows in a schematic way the developmental process that gives rise to the primordial floral organ cell–fate specification in *Arabidopsis thaliana* as predicted from a stochastic exploration of the epigenetic landscape (see next chapter for further detail). The set of attractors of the underlying transcriptional gene regulatory network correspond to the configurations characteristic of the four primordial cell types that later on during development form the four flower organs of most angiosperms (flowering species): sepals, petals, carpels, and stamens. The

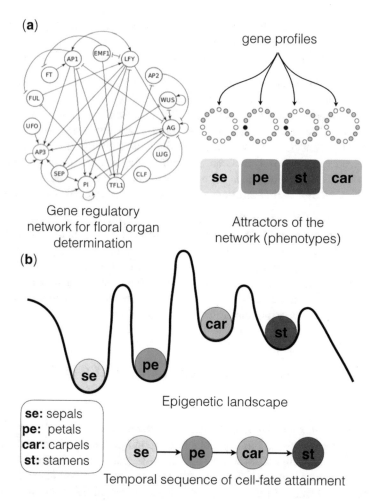

**Fig. 1.10** Epigenetic landscape. (**a**) The transcriptional gene core regulatory network that underlies primordial floral organ cell–fate determination in *Arabidopsis thaliana*. The gene profiles of the four attractors of this network (left) correspond to the cell phenotypes that characterize sepals, petals, stamens, and carpels, denoted by **se**, **pe**, **st**, and **car**, respectively. (**b**) Shows a (right) simple Waddington-like representation of the epigenetic landscape that emerges from the regulatory network. Each attractor is represented by the ball located at the bottom of the corresponding basin. The developmental trajectory that gives rise to the structure of the flower (i.e., first sepals, then petals, then stamens, and finally carpels) is also schematically represented. Computer-based simulations, performed in [18], show that this developmental trajectory can be driven by stochastic fluctuations in stochastic resonance with the underlying core gene regulatory network

floral organ primordia emerge from the population of stem cells (undifferentiated) that constitute the floral primordia.

Deterministic gene regulatory models imply that the activity of a given gene at any given time only depends the activity of the genes regulating it at a previous time step. (see next chapter). These models are useful to explore the effect of such interactions in a network in cell differentiation. But regulatory interactions are generally prone to stochastic fluctuations, and these coupled to the deterministic logical rules or interactions yield important aspects of observed biological patterns.

*Remark 1.9 (Biological Stochasticity and Its Constructive Role)* It has been proposed that in the context of nonlinear gene regulatory networks, stochastic noise may play a constructive role in biology and that present-day biological networks might have evolved in noisy conditions (see for instance [386]). Stochastic gene regulatory models consider noise into nonlinear dynamical models and such stochastic regulatory networks recover, for example, robust temporal morphogenetic patterns (See Fig. 1.10). In the biomedical context considered in this volume, such patterns may correspond to generic transitions observed both during health and disease.

The complexity of pattern formation processes in multicellular organisms is a formidable phenomenon. Gene regulatory networks collaborate in the determination of cell–fate, but pattern formation also involves the interaction of gene regulation with other complex phenomena where physicochemical fields play a fundamental role. In what follows we expose how multicellular self-organizing dynamics result from the coupled constraints that rule the interaction between informational processes and the supporting materiality of biological organisms.

## *Coupled Constraints and Self-organization in Multicellular Organisms*

During multicellular developmental processes in vivo, groups or populations of cells attain distinct fates with certain spatial and temporal patterns that occur concomitantly. Such spatiotemporal dynamics and patterns result from dynamic feedback interactions between intracellular networks and the physicochemical fields. Cell proliferation dynamics, in turn, alter such physicochemical fields that feedback to intracellular network dynamics regulating the cell-cycle and other cellular-level processes. Such processes necessarily imply multi-level modeling that requires frameworks of cooperative dynamics that simultaneously consider various levels of organization and morphogenetic patterning. Key to these multi-level models is to postulate processes that generate the positional information of cells at all times and that modulate the invariant underlying gene regulatory accordingly. Such models can be used to address issues concerning the regulation of the size and dimension of tissues as well as the relative position of organs (see for instance [17, 376]).

In addition to understanding how mechanics, geometry, and growth contribute to the formation of functional and robust structures [326], we must consider that these sources of additional constraints not only influence each other but are also coupled with at least two other fundamental dynamics coming from regulatory networks and cell proliferation. Additionally, these dynamical processes occur at different temporal and spatial scales. For example, chemical signals that are produced or excreted from cells to the extracellular matrix arrange themselves in space/time to form macroscopic patterns, which, in turn, affect gene regulatory networks in each cell, thus biasing its dynamics towards different gene expression configurations. In other words, to accomplish the extraordinary choreography implied by morphogenesis (without a choreographer!), the behavior of the chemicals or mechanic-elastic forces and communication mechanisms should be coupled to the dynamics of the gene regulatory networks in such a way that the positional cues bias the underlying corresponding attractor of the gene regulatory network, and, at the same time, the modified gene activity configuration of the network regulates the spatial pattern of chemical concentrations (see [17] for the details). We refer to the models that capture such dynamics as models of *cooperative nonlinear dynamics* [42].

Mechanical forces provide cues for heterogeneous cellular behaviors by establishing sources of positional information, thereby contributing to the regulation of morphogenesis [17, 41, 43, 189, 312, 494]. Recent work has started to uncover the molecular mechanisms by which mechanico-elastic forces are sensed by intracellular gene regulatory networks, as well as the role of myosin, actin, and tubulin fibers in cell structuring, and in the transduction of changes in mechanical and elastic forces into gene regulatory networks signals (see for instance [189, 295]). As far as human-health is concerned, the morphogenetic role of mechanic and elastic fields can be particularly relevant in the context of fibrotic diseases (see for instance [197, 282]).

As we can see, morphogenetic spatiotemporal dynamics is the result of the collaboration of gene regulation with a complex set of processes that involve constrained self-organizing phenomena. As a consequence of disrupted regulatory processes, chronic degenerative diseases also depend on this complexity. The understanding of the generic processes that underlie developmental phenomena opens the door to the systems-level understanding of human disease. The merging of:

- conceptually clear theories,
- formal computational-mathematical tools,
- and molecular-genomic data into coherent frameworks,

is at the basis of a much needed nonlinear, dynamical, systems-level explanatory and predictive approach to development (and also in fact also to evolution). We use in this volume three biomedical examples to illustrate our proposals:

- Epithelial cancer
- Chronic inflammation
- Atopic dermatitis

These biomedical examples will illustrate (see Chap. 3) not only how a bottom-up medical systems biology approach allows the systems-level understanding of the underlying disrupted regulatory processes, but also how our approach suggest potential preventive therapeutic strategies.

In what follows we shall discuss how the dynamics of human-health can be tackled by a state-space perspective.

## 1.6   Dynamical Trajectories of Human-Health

We have previously discussed that a pathology, understood as a dynamical process, can be modulated by treatment and prevention strategies (see Fig. 1.4). Moreover, we also considered through a systems-level perspective the interplay between biological development (cell differentiation, cell proliferation, and morphogenesis) and chronic degenerative diseases (characterized by health transitions that can go from the healthy state to the disease state, via the preclinical state). We also established the connection between regulatory dynamics at the transcriptional intracellular level and cellular phenotypic plasticity that underlies potential health transitions (under environmental constraints at the micro and the macro levels).

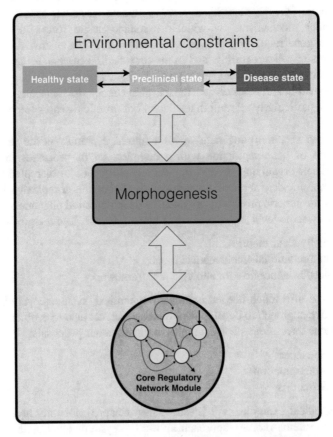

**Fig. 1.11**  Medical systems biology initial brick. This figure shows in schematic terms the interplay between biological development (i.e., cellular morphogenetic dynamics driven by transcriptional core regulatory networks) and human disease dynamics, under the dynamical constraints imposed by the environment (at the micro and the macro levels). This interplay defines the initial brick required to start the construction of the medical systems biology theoretical framework

We can summarize all this initial brick of the medical systems biology theoretical framework building in schematic terms as shown in Fig. 1.11. In what follows we show how the state-space perspective can be generalized, in order to develop health-state improvement by regulation of state-space trajectories of human-health.

## *The State-Space Perspective*

Following the conceptual framework discussed above, we can consider the systems-based description of the dynamics of human-health in terms of a state-space-based trajectory. This implies that under the selection of a conveniently multidimensional descriptive variable, i.e., the state variable (in fact a vector composed by a descriptive set of different well-characterized biomolecular activation levels or values), the evolution of the health-state of a given person can be seen as the time-evolution of the chosen set of signals that encode his/her corresponding health-state (from a clinical point of view). Therefore, the descriptive information quantifies preference-based measures of human-health. Then, we say that a given person is completely healthy (or perfectly healthy, or in good health, if we prefer) if the corresponding health-state follows a trajectory (in the chosen state-space) that displays only *clinically defined* signals or biomedical activation levels of the components being considered (necessarily related to specific clinical health indicators) that correspond to good health at any considered time.

*Remark 1.10 (Clinical Characterization of Human-Health)*  Defining human-health is not easy. For our purpose, we shall consider human-health as the well-established ability to adapt and to self-manage, under the assumption that this definition is accompanied by a set of dynamical features and dimensions that can be clinically measured. Note that the actual clinical dimension of human-health is in fact given by a trade-off between the current socio-cultural constraints and the current techno-scientific standards that define medical science at a given time. The consideration of this complex trade-off is out of the scope of this volume.

With state-health trajectories in mind, a measured disease condition will then originate a state-health trajectory that does not match the state trajectory corresponding to clinically defined healthy human dynamics. Therefore, the disturbed trajectory does not correspond to the perfect health state trajectory (we could add the adjective *unhealthy* to that undesired trajectory), and the mismatch between this undesired trajectory and a clinically defined healthy trajectory will characterize the intensity of the disease (measured in clinical terms). This implies that at a given time, the state-of-health of a given individual would then be a measurement of the distance that exists at that time between the actual health-state trajectory and the trajectory corresponding to (ideal) good health. Thus, a therapeutic intervention intended to improve the health state of the affected individual can be conceived in terms of the continuous minimization of the distance between those state trajectories.

Since regulatory networks are adaptive (which implies some level of dynamical plasticity), they can be intervened in order to modify the behavior of the associated biological system. Let us now tackle this issue.


## *Regulated Human-Health State-Space Trajectories*

Based on empirical evidence, for a given individual both the good health-state trajectory and the state trajectory corresponding to a disease condition emerge from the same regulatory networks, *working under different operational circumstances* (constrained by a complex network of multi-factorial risks, including the key modulation of lifestyle). The therapeutic interventions can then be seen as (ideally) well-designed interactions between exogenous stimuli and the concerned regulatory networks. The plasticity that characterizes regulatory networks enables that therapeutic interventions are implemented while attenuating disturbances (on the concerned regulatory networks). It is quite obvious, nonetheless, that not all therapeutic interventions are possible. This is due to the fact that regulatory networks are under structural and functional constraints, and also because of our own constraints (that include both knowledge uncertainty as well as technological limitations).

*Remark 1.11 (Technological Limitations)* We must point out that at this time the total characterization of a health-state trajectory for a given person is out of our current technological capacities. Under the socio-cultural constraints, the best we can do is to develop estimation techniques in order to get some important well-defined characteristics of that curve, and use them as a starting point to conceive dynamical therapeutic interventions. Moreover, even if we have perfect knowledge on a particular human-health issue (which is always contestable), it may be impossible to intervene in order to modify it.

Pharmacological interventions can seldom result in the total remission. Moreover, measuring a distance between good health and the actual health, at a given time, is a quite formidable task. As we shall see in the next chapters, nonetheless, at this time the state-space-based modeling approach offers some very important tools in order to advance the medical systems biology agenda in that promising direction.

It is now time to conclude this chapter. In what follows we include a brief summary, and we prepare the transition to our chapter concerning the formal framework of the modeling procedures that we shall apply later for our cases of study.

# 1.7  Synthesis

Up to this point we have argued that systems biology approaches might be very helpful to shed light on several of the current clinical research challenges posed by the increasing incidence and social and economic burden of chronic degenerative disease. Since these diseases are basically the undesired consequences of the complex interplay between harmful environmental conditions and inconvenient lifestyle processes (under the circumstances of the unavoidable human aging natural processes), with underlying developmental systems-level mechanisms, medical systems biology provides the right framework to uncover the properties of that interplay, potentiating the development of preventive therapeutic strategies (in combination with pharmacological-based approaches).

Effective prevention requires a systems-level deep understanding of disease, and the epigenetic landscape formalism allows the effective modeling proposal that integrates the theoretical and empirical exploration of the complex systems-level mechanisms that underlie the disruption of human-health. It is clear that disrupted regulation is at the heart of disease. The dynamical properties of chronic degenerative disease derive from the interplay between multi-factorial risk factors and underlying regulatory processes that are at play during normal development. Hence the complexity involved in human-health and disease is mediated by constrained biomolecular interactions that can only be understood in the light of biological developmental processes.

A bottom-up medical systems biology is required to unveil the generic processes that explain chronic diseases at a systems-level. We believe that such a bottom-up framework can be used to:

- Systematically investigate the risk of developing disease in response to different risk factor combinations.
- Develop a formal knowledge-exchange platform to interpret clinical outcomes in terms of the systems-level constraints imposed by the interaction between environmental conditions and regulatory processes.
- Elucidate the mechanisms underlying the gradual tissue deterioration that characterizes chronic degenerative diseases.
- Identify vulnerable but asymptomatic patient cohorts requiring preventive treatments.
- Design optimal treatment regimes that effectively revert the disease phenotype with the minimal associated negative side effects.

At this time, medical systems biology, supported by systems-level mathematical and computer-based modeling (nourished by the impressive biomolecular cellular-based empirical evidence that defines our post-genomic biology era), provides a mind set to tackle in a systematic manner the improvement of complex health conditions. We hope that this initial chapter has clarified our proposal.

The following methodological chapter focuses on our modeling procedures and aims at providing the formal tools that are needed to pose clinical problems in a bottom-up systems biology framework.

# Chapter 2
# Modeling Procedures

## 2.1 Introduction

Being concerned by the understanding of the mechanism underlying chronic degenerative diseases, we presented in the previous chapter the medical systems biology conceptual framework that we present for that purpose in this volume. More specifically, we argued there the clear advantages offered by a state-space perspective when applied to the systems-level description of the biomolecular machinery that regulates complex degenerative diseases. We also discussed the importance of the dynamical interplay between the risk factors and the network of interdependencies that characterizes the biochemical, cellular, and tissue-level biomolecular reactions that underlie the physiological processes in health and disease. As we pointed out in the previous chapter, the understanding of this interplay (articulated around cellular phenotypic plasticity properties, regulated by specific kinds of gene regulatory networks) is necessary if prevention is chosen as the human-health improvement strategy (potentially involving the modulation of the patient's lifestyle). In this chapter we provide the medical systems biology mathematical and computational modeling tools required for this task.

The chapter is organized as follows: We introduce in Sect. 2.2 some basic concepts on medical systems biology modeling, placing our exposition in the context of the behavior of complex systems (recalling important concepts like *complexity* and *emergence*). We focus our attention on the key role played by feedback-based interactions in determining of the dynamical properties of biological networks.

In Sect. 2.3 we deal with the practical definition of *system* and *modeling*, and we expose how the experimental context conditions systems modeling. We also discuss why medical systems biology requires systems-based mathematical modeling, and present a brief classification of mathematical models (taking into account the role played by the unavoidable presence of uncertainty). Additionally,

© Springer International Publishing AG, part of Springer Nature 2018
M. E. Álvarez-Buylla Roces et al., *Modeling Methods for Medical Systems Biology*,
Advances in Experimental Medicine and Biology 1069,
https://doi.org/10.1007/978-3-319-89354-9_2

we include the general mathematical modeling procedure, explain the particularities of the interplay between modeling and simulation, and justify why we prefer to work with models that allow hypothesis verification. We conclude the section with exposition of the conceptual framework that characterizes the state-space description of dynamical systems.

Section 2.4 is dedicated to justifying the use of a mechanistic bottom-up approach based on dynamical systems theory to tackle the description of biological systems (defined there as complex networks of biomolecular interactions at the intracellular level). We also explain the systems-level consequences of the modular structure of complex biological networks. As modeling is concerned, we define systems biology in bottom-up terms, and introduce a practical definition of *biological mechanism*. We also argue how to guide the modeling process through the consideration of the regulatory dynamics that underlie the behavior of biological systems (taking into account transcriptional dynamics), and expose the connection between disease and disrupted regulation, and how this phenomenon can be understood through systems modeling (taking into consideration that cells satisfy the requirements to be considered dynamical systems). We conclude the section exposing the natural connection between cell phenotypic plasticity and the dynamical systems state-space perspective (the dynamical systems concept of attractor is chosen to describe observable cellular phenotypes).

Taking into account the behavior of single cells, we present the modeling tools required to tackle its description in Sect. 2.5, following this with a deterministic discrete-time and discrete-space perspective. The discrete Boolean approach is then discussed, specifying the essential components of gene regulatory Boolean networks, namely the *set of genes* that constitute the nodes of the network (and the members of the state-vector), and the *updating rules* that characterize how the activity state of each given node in the network is regulated by the activity state of the whole set of nodes in the network. We also tackle the dynamical analysis of Boolean networks, which in the context of medical systems biology allow us to describe cellular phenotypic identities in terms of the steady states (i.e., the attractors of the system) of a given gene regulatory Boolean network. The computer-based methodology to test the consistency between the model's predicted phenotypes (due to mutations) and the observed cellular phenotypes is also discussed.

Section 2.6 is concerned by the continuous approximation of discrete Boolean dynamics. This approximation allows us to build a computer-based methodology intended to explore the role of specific genes in transient dynamics. We also consider how to use Boolean networks to explore the phenotypic consequences of gene decay rates. Transient dynamics are also tackled in Sect. 2.7 by introducing a stochastic methodology intended to study gene regulatory networks and phenotypic plasticity. The proposed stochastic formalism supports the development of the epigenetic landscape theoretical framework, which makes it possible to uncover in gene regulatory networks transient dynamics associated to the structure of the stochastic epigenetic landscape. Specifically, the methodology allows the estimation of transition probabilities of attractors (which can then be applied to uncover normal and disrupted developmental paths).

Since the discrete Boolean approach is qualitative, it is not suitable to explore the effects of specific biochemical mechanisms and of quantitative variations. The aim of Sect. 2.8 is to deal with this issue via the application of mechanistic continuous models to the description of medical systems biology phenomena. These models take the form of systems of nonlinear ordinary differential equations. From the fact that cells commit to a phenotype through nonlinear signal processing of microenvironmental conditions by regulatory networks, continuous modeling allows an understanding of the key role of feedback-based interactions as well as parameter variations. The section provides a useful methodology to build and analyze mechanistic nonlinear ordinary differential equation models from scratch (including the identification of initial conditions and the parameters of the system, under the constraints imposed by the empirical evidence). We also expose in the section how to assess the robustness/plasticity behavior of continuous models in response to perturbations, and consider multi-scale dynamics in the context of the interplay between regulatory networks and the microenvironment (this allow us to tackle important questions turning around tissue-level consequences of disrupted phenotypic dynamics).

We conclude this chapter in Sect. 2.10, discussing how to apply the exposed modeling methodologies to shape an exploratory protocol intended to elaborate predictive hypothesis (which opens the door to develop a research agenda focused on preventive strategies to deal with chronic degenerative diseases).

## 2.2  Basic Concepts of Medical Systems Biology Modeling

### *Complexity and Emergence*

Science is about organized knowledge, and systems modeling is about systems-based methodologies for the efficient production of meaningful knowledge. Thus, no scientific agenda concerned by complex systems (e.g., medical systems) can exist without goal-oriented systems modeling. Therefore, we decided to begin our exposition on modeling procedures tackling this topic. So, in this section we follow a quite utilitarian perspective to expose the basic modeling issues that we consider are required in the context of *medical systems biology*. We must point out that by following this utilitarian approach we voluntarily decide to be at the same time formal and to avoid excessive mathematical abstraction. Then, we prioritize the apprehension of practice-oriented conceptual tools (by biomedical researchers concerned by human health issues). With this, we also explore the standardization of the basic concepts that currently flow in the available medical systems biology scientific literature, where a lack of consensus still persists. Throughout our exposition we refer to useful bibliographical sources that deeper information on the discussed modeling procedures proposed here.

To proceed with our exposition, let us first include here a brief discussion concerning the important concept of *complexity*. This is important because when tackling medical problems, from a medical systems biology point of view, we deal in fact with *complex systems*.

**Definition 2.1 (Emergence)** Related to the behavior of dynamical systems, it consists of the process of coming into existence.

We are conscious that this definition can be judged as being ambiguous. And such judgment would be justified. Therefore, to be more specific, the notion of emergence that concerns us here is that of *weak emergence*, developed in the context of the study of physical systems (see for instance [447]). This type of emergence is well suited for medical systems biology and in fact corresponds to the systems-level emergent property that is amenable to computer simulation (for an interesting discussion on the meaning of weak and strong emergence, see [79]).

*Remark 2.1 (Holistic vision)* In the context of the study of dynamical systems, understood as the well-characterized composition of its constitutive parts, a given detected behavior is named emergent when it is not a property of any of the components of the system, but rather results from the system's components interactions. Therefore, the quest to understand of emergent behaviors requires a holistic perspective. Note that this notion implies that dynamical systems can be modeled as networks, that is, organized collections of interacting components.

Contemporary science uses the term "emergence" to refer to the manifestation of novel collective phenomena or patterns in some large systems stemming from a complex organization of their many constituent and interacting parts or components. Thus, emergence and complexity are inseparable concepts. So:

**Definition 2.2 (Complexity)** A given dynamical system is said to be complex when it is composed of a collection of individual agents whose interactions give rise to emergent behaviors.

This definition implies the existence of dynamical systems that are not complex, that is, they do not display emergent properties at all (the behavior of such a system can be totally understood by the analysis of its constitutive parts).

Note that a given individual agent belonging to a system might be a system (even a complex system) by itself.

To illustrate the previous concepts, let us consider now some examples of emergent properties displayed by some dynamical systems.

**Cell self-sustainability:** A living cell is a rich collection of organized protein machines. Each biomolecular complex in the family collaborates with its fellows to manage the flux of matter and energy that ensures the survival of the cell. A system is self-sustainable if it can maintain itself independently of other systems. This is one of the most important properties of any living cell, which emerges from the interaction of the cell's constitutive biomolecules as well as chemical and physical fields.

**Population density:** A given population is defined as a measurement of population per unit area or unit volume. This is an emergent trait of the whole population that no single individual could exhibit on its own.

**Universal health care:** We say that a country has an universal healthcare system if that system provides health care and an associated financial protection to all its citizens. As an emergent property, universal health care comes as a result of the interaction of a diverse collection of interacting systems that includes, among others, a tax revenue system, a health insurance system, a fund system, and a political system.

As illustrated by these examples, emergent properties are the result of the system's components interactions. In this context, feedback-based interactions are particularly meaningful.

### → Feedback-Based Interactions

As can be seen, the notion of complexity is intimately related to the study of the behavioral consequences of the interactions between the constitutive components of systems. It should be mentioned that of the different types of interactions, those that are based on feedback are by far the most interesting and the most important ones to understand systems-level behavior and phenomena. By feedback we mean that if a cause, say $\mathbb{C}$, lead to an effect $\mathbb{E}$, $\mathbb{E}$ will also affect $\mathbb{C}$ (this is also called *circular causality*). Feedback-based interactions are related to systems-level properties:

- Multistability
- Homeostasis
- Robustness
- Plasticity

These are precisely the kind of properties that characterize medical systems. In fact, feedback-based interactions are ubiquitous in complex systems. In general feedback-based interactions have two flavors (see Fig. 2.1):

**Negative feedback:** This kind of interaction gives rise to a flow of information that forms a closed loop that connects the output of a given system with its input, having as its characteristic dynamical consequence the reduction of fluctuations in the system's output (which, under some conditions, may promote system stability).

**Positive feedback:** This kind of interaction gives rise to a flow of information that forms a closed loop that connects the output of a given system with its input, promoting the reinforcement of the signals flowing in the loop. Under some conditions, this reinforcement may promote system instability.

Colloquially speaking, we can say that negative feedback interactions (under some circumstances) promote equilibrium or homeostasis, whereas positive feedback interactions (in coordination with negative feedback-based interactions) promote multistability. Nonetheless, in biological systems this sign separation (negative

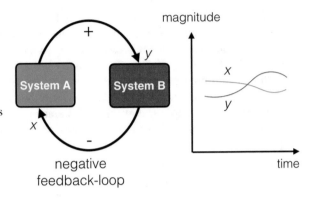

**Fig. 2.1** Feedback-based interaction. The output of System A is routed back as a processed input (by System B). This chain of cause-and-effect forms then what is called a feedback loop. Negative feedback tends to reduce the fluctuations in the output (promoting equilibrium). In contrast, positive feedback promotes reinforcement of the signals flowing in the loop

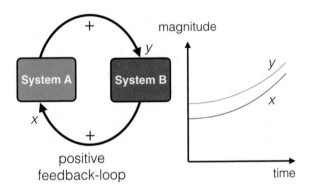

or positive) is not always so clear. A given feedback-based biomolecular inter-action might be positive or negative, depending on the given circumstances. Nevertheless, structural and functional constraints frequently determine the sign of feedback-based interactions. In general, in biological networks it is common to find interactions that combine negative feedback loops with positive feedback loops. Such combinations are critical for the emergence of the dynamical properties of biological networks (Fig. 2.2).

*Remark 2.2 (Self-organization and Self-catalysis)* As an evolutionary result, the topology of biological networks reflect the interplay between self-catalyzed and self-organized biomolecular processes. This interplay, which rules the evolution of open thermodynamical systems, underlies the formation of complex biomolecular systems (see for instance [34, 153, 173]). Negative and positive feedback-based interactions play a key role in the structural and functional topology of such kinds of dynamical systems.

As far as medical systems are concerned, complexity is a systems-level charac-teristic that defines them at all levels of description, from the biomolecular scale to the organization of medical institutions (see for instance [373]). The complex

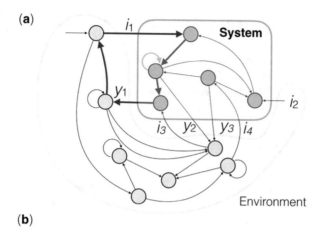

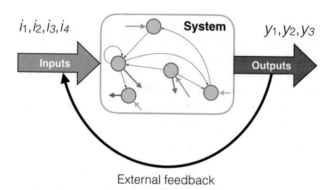

**Fig. 2.2** Dynamical systems as networks. (**a**) In abstract terms, a dynamical system is constituted by a (goal-oriented) collection of interacting components (shown in green), that is, a dynamical network, that is linked to its environment through input and output channels. For the represented system, $\{i_1, i_2, i_3, i_4\}$ and $\{y_1, y_2, y_3\}$ denote the vector of inputs and outputs, respectively. The system's components could be interacting between them in different ways, including via feedback-based interactions (as indicated by the purple and the orange edges that in graph theory terms constitute cycles). We also represent the fact that a non-isolated system interact with its surroundings. The immediate environment can or cannot be known in detail. (**b**) Shows the chosen system as well as the interaction with its environment. The system is represented in input–output terms. The feedback-based interactions that link the system's outputs with the system's inputs are represented via the shown exogenous feedback connection. The arrows in blue and in red represent how inputs and outputs are related to specific components of the system

nature of medical systems asks for strategic thinking to ease understanding. Thus, formalization and abstraction are cognitive tools that are required for this. We have been using until now an intuitive notion of system; let us now formalize in what follows the concepts of system and modeling, and to expose how experiments are related to these key concepts.

## 2.3   Systems, Modeling, and the Experimental Context

In general, it is obvious that the complexity of our world is overwhelming. This implies that it is very hard to get scientific explanations for the complex phenomena that shape our world. That is the reason why science tackles the understanding of the world through:

- Systems-level modeling
- Knowledge abstraction
- Systems reduction

We can say that we confront complexity through modeling, making it cognitively tractable. This goal-oriented cognitive tool can be defined as follows:

**Definition 2.3 (Modeling)** The cognitive activity that consists of thinking about and making goal-oriented representations to describe how particular systems behave.

Thus, to model a given system is to organize the knowledge that concerns it.

Therefore, the cognitive reduction of complexity leads us to the identification of physical reality as composed of a huge set of well-defined interacting systems. A given system can then be represented by a specific model. The chosen description organizes knowledge in a purposeful, specific way.

From a practical point of view, systems can be defined as *processors of information*. This leads us to the following [77]:

**Definition 2.4 (System)** A potential source of data.

This utilitarian definition implies that a system can provide information, *if it is requested to do so*. This implies that information from a given system is collected through an interactive process. For a given system, the collected data are the result of a goal-oriented knowledge harvesting process that can include both observation of the behavior of the system as well as the conscious, purposeful manipulation of it. In the latter case, a stimulation process of the system, that is, an experiment being carried out on the system, can extract useful information that can be used to uncover the specific mechanisms underlying the information processing that explains what the system actually does. This leads us to the following:

**Definition 2.5 (Experiment)** Goal-oriented stimulation of a concerned given system, in order to get essential and useful information related to it, intended to provide the raw material that is required to build goal-oriented models.

*Remark 2.3 (The Model and the Experiment as an Inseparable Couple)* A particular model of a given system cannot be separated from the experiment that provides the data (i.e., the empirical evidence) that give rise to the organized knowledge coded by the model. The experimental context defines and validates the chosen model.

It is very common to start a modeling task with preexisting empirical data, that is, experiments are not always carried out as a constitutive part of the modeling process. This is in fact a common situation as far as contemporary medical biomolecular systems are concerned. Indeed, it is quite common nowadays to have more data than corresponding explanations turning around its biomedical meaning. Moreover, the pace of biological data generation by high-throughput technologies shows super-exponential growth; see for instance [180, 368]. Medical systems biology needs to take this situation into account and put efforts in integrating and analyzing this data.

Abstraction and formalization of knowledge, in order to choose useful models of specific systems, are strengthened when they are coded in an effective manner. As a modeling language, mathematics offers a very effective tool to code and to share organized knowledge. Let us now discuss this topic.

## *Mathematical Modeling*

In practical terms, knowledge is useful when expressed through quantities that can be processed via workable computations (intended to uncover systems-level basic mechanisms hidden in the processed data). Hence, we are interested in qualitative and quantitative descriptions of dynamical systems (i.e., systems that explicitly depend on the time variable). Henceforth, we shall code our organized knowledge extracted from dynamical systems using the standardized language of mathematics. We are then concerned by mathematical models of dynamical systems:

**Definition 2.6 (Mathematical Model)** Describes relationships and variables that offer a formal explanation of the data extracted from the study of the concerned dynamical system. The identified explicative relationships describe the causal mechanisms involved in the processing of information, while the proposed variables usually represent, in quantitative terms, the signals being processed by the modeled system.

It is not an exaggeration to say that the coupling of experimentation and mathematical systems modeling is a fundamental part of the backbone of a broad spectrum of scientific endeavors. In fact, mathematical modeling is essential for science. Moreover, because each particular scientific field has its own particular needs, mathematical models exist in diverse tastes, each one depending on the specific goal pursued by a particular model, as well as the particularities of the systems of interest of each particular scientific area.

Let us now include here a brief classification of mathematical models, which is by no means unique.

→ **Classification of Mathematical Models**

A mathematical model is composed by interacting operators, each one representing an information processor. In terms of the effects that the constitutive operators produce in the incoming signals, we can classify mathematical models as:

**Linear mathematical models:** We say that a mathematical model is linear when all the involved constitutive information processors only perform time-independent proportional changes on the incoming signals. In other words, any processed signal by a given processor is just a proportional amplification or attenuation of the corresponding input signal. Moreover, for a linear mathematical model the consequences of the action of a combined set of stimuli equal the combined effects of the set of stimuli taken separately. This is known as the *superposition property* of linear systems, which is very useful when studying the interaction of linear systems with their surroundings. When a given dynamical system satisfies these two properties, we refer to it as a linear system.

**Nonlinear mathematical models:** A mathematical model is considered to be nonlinear if at least one of the involved constitutive processors does not act in a linear way on its corresponding incoming signals. An incoming signal and the corresponding output signal are not just proportional one to the other, and consequently the separation property is not valid in this case.

Remember that we are interested in mathematical models as useful representations of dynamical systems. This class of systems explicitly depends on the time variable. Therefore, the way in which this independent variable is quantified establishes a direct classification of mathematical models as follows:

**Continuous-time mathematical models:** If the time variable can take any value between two given time instants, we say that corresponding mathematical model evolves in a continuous-time manner.

**Discrete-time mathematical models:** When the time variable takes its values in a well-defined set composed of distinct separated points of time.

In general we consider here non-autonomous systems. So, we consider that the dependent variables explicitly depend on the independent variable, that is, the time variable.

*Remark 2.4 (Qualitative Descriptions and Continuous-Space Models)* When the dependent variables of a discrete-time model are discretized, we say that the model is qualitative. These kinds of models are usually coded by finite state representations. For instance, binary Boolean networks code systems in a qualitative manner. If the dependent variables of a given model are not discretized, we deal with a continuous-space model.

Not all discrete-time models are also discrete-space models. Indeed, the dependent variables of discrete-time models coded in terms of *difference* equations are usually not discretized. We are not considering these kinds of models in this volume.

In this volume we consider continuous-space models when working with continuous-time models, and discrete-space models when working with discrete-time models. To simplify our exposition, in what follows continuous-time mathematical models just called continuous models and discrete-time mathematical models just called discrete models.

As a goal-oriented cognitive task, mathematical modeling is constrained by the unavoidable limitations that characterize any knowledge recollection process. Modeling suffers always from uncertainty.

## → Uncertainty and the Lack of Predictability

Uncertainty is intrinsic to the acquisition of scientific knowledge, and its unavoidable presence in the context of systems modeling implies that perfect representation of dynamical systems is in fact unachievable (in fact, only a perfect copy of a given system would correspond to a perfect model of it). Henceforth, mathematical models are imperfect. Uncertainty can be the result of limitations of the involved measurement processes (i.e., measurement noise) or it can be the result of unavailable information related to the intrinsic nature of the given system. We must point out that for a given system, and for a given modeling goal, it is important to look after the best possible model, and take into account for this the presence of uncertainty, which can be due to:

- Unmodeled dynamics
- Exogenous disturbances
- Noise

Moreover, it is always a matter of choice to include or not in a mathematical description of a given system a model of the involved uncertainty (depending on the concerned goal). It is common to model uncertainty in terms of lack of predictability of the behavior of the system under certain operational conditions. Moreover, predictability can or cannot be related to randomness. For our purpose, *we shall consider in the sequel that the lack of predictability is due to randomness.* We must point out that randomness can be due to the presence of measurement noise or can originate from the system's intrinsic nature (or can result from the presence of both phenomena). Taking into account the predictability issue, we consider then the following model classification:

**Deterministic mathematical models:**   When the mathematical description does not include any consideration on the lack of predictability (i.e., the variable states are described by unique values).

**Stochastic mathematical models:**   When the values of the involved variables are described by probability distributions.

We shall see later that the inclusion of stochasticity in biomolecular network modeling has revealed the importance of so-called stochastic resonance phenomena that result from the combined action of random fluctuations and deterministic dynamical behavior. Such stochastic resonance seems to explain some apparently deterministic behaviors and patterns in biological systems (see ahead).

*Remark 2.5 (Stochasticity in Biomolecular Systems)* Cell biomolecular systems involve interactions of finite sets of molecules. These interactions take place in an aqueous solution on the micro- and nanometer scale and thus are subject to thermal fluctuations, which necessarily give rise to a form of stochasticity known as intrinsic noise. This property implies that the behavior of cell biomolecular systems can be considered basically as deterministic, but under the influence of random noise. The frequential nature of the present noise, as well as its strength, depend on specific spatiotemporal circumstances. Another source of random fluctuations also results from the fact that the number of molecules involved in any interaction is finite and sampling stochasticity is then involved.

### → **Multiplicity of Representations**

We must insist that for a given system the choice of a particular mathematical description is a goal-oriented process. This means that different goals very often imply the selection of different models (it is quite uncommon to choose the same model for different modeling goals). Therefore, the same dynamical system can be described by a linear or nonlinear mathematical model, for instance, or the description can consider that the time evolves in a continuous way or in a discrete manner. Moreover, complex physical systems can display behaviors that require to be captured by hybrid mathematical models, that is, models that include components that belong to more than one class of mathematical models (e.g., some part of the model can be deterministic and some other part can be stochastic). A modeling rule of thumb is to make the model to be structurally and phenomenologically similar to the described system, and take into consideration the limitations imposed to the analysis of the model by the available formal and computational tools.

*Remark 2.6 (Hybrid Models)* In general, a given system is constituted by the interconnection of a given set of well-characterized subsystems. In mathematical modeling terms, the description of each one of these subsystems could require a particular goal-oriented mathematical model, since each subsystem may be related to different modeling purposes. Therefore, the whole mathematical model for the given system could be composed by a set of interconnected mathematical models, each one of them belonging to a particular class of mathematical models. In the case that the chosen model involves more than one model class, we say that the concerned system is being described in hybrid terms or with a hybrid mathematical model. Under some particular circumstances, hybrid models can be very useful. However, hybrid models can be more difficult to analyze than non-hybrid models. In mathematical terms it is not easy to analytically sustain the coherence of the modeling tools when different types of modeling approaches are combined into a single model. Therefore, it would be preferable to work with non-hybrid models, but unfortunately it is not always possible. In several types of biological cases,

**Fig. 2.3** Basic criteria for the classification of mathematical models. A given mathematical model, chosen for the description of a given dynamical system, possesses properties that are determined by proportionality to exogenous stimuli, evolution of the considered variables over time, and dependent-chance behavior. This system's characterizing interplay can be used as a conceptual framework to classify mathematical models

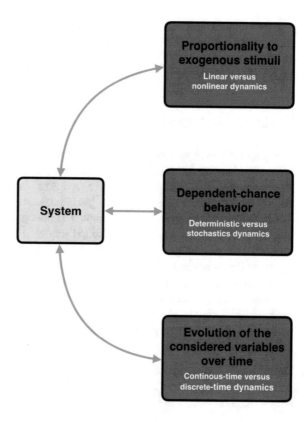

hybrid models are necessary. For example, complex intracellular gene regulatory networks can be represented by discrete models, but the interaction of the cell with microenvironmental signals may require a continuous approach.

The classification provided above is common in the modeling scientific literature. This classification is based on how the chosen model captures the given dynamical interplay between three fundamental aspects characterizing systems behavior (see Fig. 2.3):

1. Proportionality to present stimuli.
2. Evolution of the considered variables over time.
3. Involvement of stochasticity.

This classification of models is not the only available mathematical modeling classification, but it is a quite useful one.

In a given particular case, the model chosen in order to describe a specific given system would be defined by the specificity of the interplay of the described system with the alternative frameworks that define each one of the three fundamental aspects of a system's behavior (proportionality to exogenous stimuli, dependent-chance behavior, and evolution over time of the considered descriptive variables, as shown in Fig. 2.3).

Please consider [167] as a good reference to tackle the understanding of the nature of mathematical modeling in general.

### → The General Mathematical Modeling Procedure

The general mathematical modeling process of dynamical systems (biological or not), is summarized in schematic terms in Fig. 2.4. This process defines an algorithm, which is iterative by nature (see Algorithm 1), and it is continuously shaped by the modeling goals. These goals possess a certain degree of flexibility, that is, the modeling goals must be adapted to the constraints imposed by the given modeling task. Therefore, the right (ideal) model is the one that provides optimal user satisfaction under the constraints imposed by experimental accessibility to the system under consideration, as well as the available conceptual and modeling tools. In any case, the optimal type of model to be used in each case can be adapted along the way. In systems dynamical terms, a preliminary evaluation of the quality of the chosen model, that is, the preliminary evaluation of its actual *usability* value, must be based on a well-posed process of simulation. This consideration takes us to the following:

**Definition 2.7 (Simulation)** In terms of a chosen model for a given system, simulation is understood as virtual experimentation (i.e., experimentation carried out on the model and not on the described system itself).

The purpose of simulation is to detect the strengths and the weaknesses of the model as a useful descriptor and predictor of the represented system. Moreover, the experimental context establishes the validity of the model as well as the possibilities and constraints of simulations.

Once corrected (if required), after the preliminary evaluation, the model should address some well-chosen hypothesis on the behavior of the described system under stimuli, first tested on the model being considered (see Fig. 2.4) and Algorithm 1. However, not all the mathematical models are necessarily oriented to hypothesis verification. But the most useful models to understand biological systems-level mechanisms are those that are useful to verify research hypotheses derived from experimental approaches.

*Remark 2.7 (Models and Hypothesis Verification)* It is not an exaggeration to say that models that do not allow hypothesis verification should be discarded, since they have a limited usability.

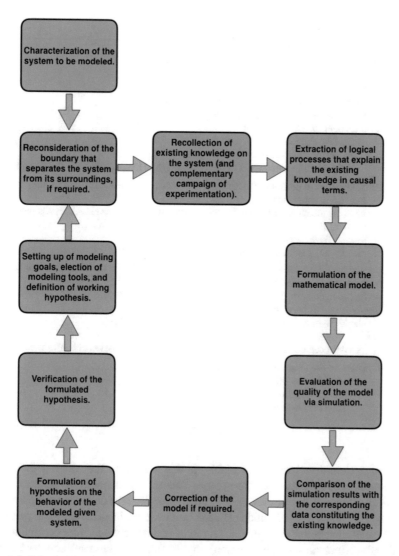

**Fig. 2.4** Schematic representation of the overall modeling process. This flowchart summarizes the logic that underlies the modeling process. As can be seen, this process has an iterative nature shaped by the modeling goals. The process of modeling begins once a specific system (to be modeled) has been chosen, and the iterative process is halted when the goals of modeling are satisfied (and restarted when necessary)

### → Recycling Knowledge and the Standardization of Mathematical Modeling

Given a specific dynamical system that requires to be modeled in mathematical terms, it is common to start the goal-oriented modeling task of recycling existing models. Thus, in the mathematical modeling process, the recollected existing

---

**Algorithm 1:** Mathematical modeling procedure

---
**Data**: characterization of the system to be modeled
**Result**: mathematical model that satisfies the modeling goal
initialization;
**while** *not satisfied with the model* **do**
> set boundaries between the chosen system and its surroundings;
> recollect the system's existing knowledge;
> **if** *experiments are required* **then**
>> define the experiments' purposes;
>> perform the experiments;
>> add the experimental results to the data collection that characterizes the system;
>
> extract the logical processes that explain the existing knowledge in causal terms;
> formulate a mathematical model;
> simulate the model to validate it through comparisons of the simulation results with the data collection;
> **if** *the model requires to be corrected* **then**
>> modify the model in order to satisfy the modeling goals;
>
> explore the model and formulate from the results of the exploration hypothesis on the behavior of the modeled given system;
> perform a hypothesis procedure;
> **if** *the hypothesis is verified* **then**
>> label the model as oriented to the verification of hypothesis ;
>
> **else**
>> set a procedure intended to uncover why the hypothesis has not been verified;
>
> **if** *the model does not satisfy the modeling goals* **then**
>> redefine the modeling goals;
>> update the selection of the modeling tools;
>> redefine the working hypothesis;

---

knowledge should include the available models related to the concerned system. Those models could correspond to systems that are in fact related to the concerned system in structural terms, even if they belong to different domains of the system. Moreover, a mathematical model should be designed in order to be recyclable, which means that both the mathematical model and its corresponding computational codification (including documentation) should follow (industrial) standardization rules. Such standardization eases the communication of knowledge. As discussed in [218], the Systems Biology Markup Language (SBML) initiative illustrates how standardization is required in order to improve the efficiency of systems biology computer-based mathematical modeling.

*Remark 2.8 (Models as Containers of Purposeful Knowledge)* Mathematical and computer-based models can not only be conceived as descriptors of specific systems, but also as symbolic containers of knowledge related to the underlying modeling goals.

We can at this level introduce the state-based approach for the description of dynamical systems.

## *The State-Based Description*

To describe a dynamical system the first step is to identify the relevant variables and their interactions with the surroundings. Next, it is compulsory to choose the descriptive variables. As inferred by its qualifying adjective, the descriptive variables describe the temporal behavior of the system under consideration. The *state-based* approach offers a well-established modeling methodology for this purpose. In formal terms, the state-space approach (also known as *state-space theory*) deals with dynamical models describing both the internal dynamics of the physical process under consideration and the interaction of this process with the outside world. The formal mathematical treatment of dynamical systems from the state-space approach of dynamical systems is beyond the scope of this book (to develop a deep understanding of this subject consider the seminal book [196]). But for the sake of clarity we shall introduce some of the basic concepts underlying this theoretical approach for the description of dynamical systems.

The state-space approach (for the description of dynamical systems) is based on a general algebraic *formal* concept of dynamical systems theory. This *algebraic* concept requires some main terms:

- Time domain
- External variables
- Internal state
- State transition map
- Output map

These terms are usually defined as follows (see for instance [196]):

**Time domain:**  Since any dynamical system evolves in time (the independent variable), its behavior is best described in mathematical terms using specific time-dependent functions. This means that every dynamical system has an associated time domain, represented in symbolic terms by $T \subset \mathbb{R}$ (with $\mathbb{R}$ standing for the field of positive real numbers). This term may be continuous, that is, $T$ is an interval (e.g., $T = [0, \infty)$) or discrete, that is, $T$ consists of isolated points in $\mathbb{R}$ (e.g., $T = \mathbb{Z}$, the set of all positive integers).

**External variables:**  As we have been pointing out, the dynamical systems that concern us are those that interact with the exterior world. In mathematical modeling terms, this interaction is coded by a set of variables that is divided in two subsets (see Fig. 2.5a): the set of inputs and the set of outputs. The given set of inputs defines a time-dependent input vector, say $u \in (u_i)$, and the set of outputs defines an output vector, say $y \in (y_j)$ (the positive integers $i$ and $j$ index a specific input-signal channel and a specific output-signal channel, respectively). The set of inputs includes two different types:

- Manipulable inputs
- Non-manipulable inputs

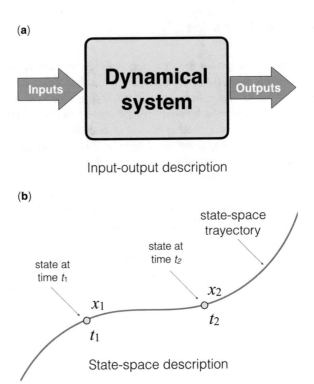

**Fig. 2.5** State-space representation of dynamical systems. (**a**) Schematic representation of a non-isolated dynamical system, characterized by the interaction of the system with its surroundings. The interaction takes place as exchange of information, coded as (manipulable and non-manipulable) input and output (regulated and measured) signals. If the system is described in input–output terms, prioritizing in the description the dynamical interplay between the environment and the system, we are concerned by input–output descriptions of dynamical systems. (**b**) Schematic representation of a system's state-description, where the internal behavior is described by the state variable. The system's state codes the interplay between the exogenous inputs and the continuously updated memory of the system

The first type includes the inputs that can be controlled by the *experimenter* (i.e., the signals that act as manipulable inputs can be *designed*). The second type includes exogenous inputs that cannot be controlled, and are commonly known as *disturbances* (which means that they act in an undesirable way on the affected system). Therefore, the definition of a dynamical system needs the specification of a set of input values, say $U$, and a set of output values, say $Y$ (i.e., in terms of mathematical operators terminology $u(\cdot) : T \rightarrow U$, and $y(\cdot) : T \rightarrow Y$). Since input and output functions cannot be arbitrary, the specification of the sets of input and output values also requires the inclusion of admissible sets, i.e., $\mathscr{U} \in U^{T}$, and $\mathscr{Y} \subset Y$, for the admissible input functions, and the admissible set of output functions, respectively.

**Internal state:**   This is the key notion in the definition of a dynamical system (in state-space formal terms), and concerns variables that describe the processes in the interior of the system. To be considered a state vector, a given set of internal variables must satisfy three conditions:

1. The present state and the present input vector determine the future states of the system under consideration.
2. Given an initial state at some initial time, the state at any later time only depends on the input values for the interval starting at the initial time and the present time. Moreover, knowledge of the initial state at some initial time supersedes the information about all previous input and state values.
3. The output value at a given time is completely determined by the simultaneous input and state values, i.e., the past inputs act on the present output only via accumulated effects on the system's present state.

**State transition map:**   The evolution in time of a state trajectory can be described by a functional map called the state transition map, say:

$$x(t) = \varphi\left(t; t_0, x^0, u(\cdot)\right), \quad t \in T_{t_0, x^0, u(\cdot)},$$

where: $\varphi$ represents the state transition map; $t_0$ denotes the initial time; $x^0$ denotes the initial state, i.e., the state of the system at time $t_0$; $T_{t_0, x^0, u(\cdot)}$ denotes the time interval of $T$ starting at $t_0$, and being characterized by the presence of the stimuli vector $u(\cdot)$. Note that the initial condition and the input condition determine the state trajectory through the transition map.

**Output map:**   In agreement with the concept of internal state, the definition of a dynamical system requires the inclusion of an output map. That operator maps the state of the system on the output vector (i.e., fixes the transmission of information from the system to its surroundings). This map, say $\eta$, is conditioned to be determined by the state and input values at time $t$:

$$y(t) = \eta(t, x(t), u(t)).$$

Now, using the state-space approach, we can introduce the axiomatic definition of dynamical system as follows (see [196]):

**Definition 2.8 (Dynamical System)**   An algebraic structure $(T, U, \mathscr{U}, X, Y, \varphi, \eta)$ is said to be a dynamical system or state-space system with time domain $T$, input value space $U$, input function space $\mathscr{U}$, state-space $X$, output value space $Y$, state transition $\varphi$ and output map $\eta$, if $T, U, \mathscr{U}, X, Y$ are non void sets, $T \subset \mathbb{R}$, $\mathscr{U} \subset U^T$, and $\eta : T \times X \times U \to Y, \varphi : \mathscr{D}_\varphi \to X$ (where $\mathscr{D}_\varphi \subset T^2 \times X \times \mathscr{U}$) are functions such that the following axioms hold:

**Interval axiom:**   For every initial time, initial state, and for a given control input, the life span of the corresponding transition map is an interval in $T$ containing the initial time.

**Consistency axiom:**    For every initial time, initial state, and control input, the
transition map maps these data on the initial state.
**Causality axiom:**    For all initial time, and for a given initial state, if two control
signals are identical in a time interval starting in the same initial time, the states
mapped at the end of the interval coincide.
**Cocycle property:**    If a given time instant, say $t_1$, belongs to the life span of the
transition map characterized for some initial time, some given initial state, and
some given control input, with a corresponding state resulting from the action of
the transition map, then the time interval defining the life span of the transition
map having $t_1$ as its initial time is contained in the life span (of the transition
map) characterized by the initial time. Moreover, for the life span of the transition
map starting at $t_1$, the corresponding state trajectory matches the state trajectory
evolving from the initial time.

This axiomatic definition has a formal statement that is beyond our scope; for the
details see [196].

Notice that the previous definition considers time a continuous variable. But
the same definition can be easily adapted to consider the discrete-time case (or
even a hybrid case, if dynamics in both continuous-time and discrete-time are
simultaneously taken into consideration for modeling purposes).

*Remark 2.9 (State-Space Theory Describes Systems in Terms of Information Storage)* The previous axiomatic definition implies, in colloquial terms, that the state
of a dynamical system is a kind of continually updated memory or information
storage (causality is embedded in the definition). Thus, the state variables need not
to represent physical quantities (of course, we would always prefer to choose state
variables that are physically meaningful!). However, in order to avoid arbitrariness
in the selection of the state variables, it is quite common to choose them in order to
represent the minimal amount of information required to describe the effect of past
history on the future development of the system (for specific initial conditions as
well as for specific input stimuli).

In the context of this volume, we choose the state-based approach for the
description of the dynamical systems that concern us, because of the conceptual
advantages offered by this formalization. With this approach we can code our
knowledge on medical dynamical systems through equations (and computational
procedures), and from them we can conceive empirically validated methodologies
to solve medical issues. In what follows we shall apply the conceptual framework
of the state-based approach to tackle biomedical systems.

It is now time to tackle biology from a systems dynamics point of view.

## 2.4   Systems Biology

### *Basic Concepts*

From a scientific point of view, life is a matter of biological dynamical systems. As we have seen, in general any system (biological or not) can be seen as just a potential source of data. The data, which can be collected through empirical exploration, provide the raw material for the construction (or selection) of requested models.

Biological systems are composed of relevant biological entities. Relevancy is related to the level of detail considered in the modeling process, i.e., it depends on the chosen granularity of the model.

Since biological systems are extremely complex, the granularity of the mathematical model chosen to describe a given biological system is quite important. In our case, we are interested in biomolecular dynamics at the cellular level related to a particular class of phenomena: the transition from a healthy to a disease state of the system under consideration. That is then our granularity election, justified because we want to uncover mechanistic basic principles underlying human disease. We postulate that some of the basic aspects of such systems-level mechanisms originate at the level of the complex biomolecular networks within cells and their interactions with physical and chemical fields, as well as with signals in the cellular microenvironment. Taking into account this choice, we shall consider here the following:

**Definition 2.9 (Biological Systems)**  A complex network of interactions between biological entities, such as biomolecules, cells, tissues, organisms, populations and communities.

*Remark 2.10 (Focus on Biomolecular Systems)*  There are many types of biological systems. However, this volume forces on biological subterms operating at the level of molecular biology, since it is mainly biomolecular dynamics what underlies the dynamics of chronic degenerative diseases.

As was pointed out before, when we talk about complex systems, we are following the current essential scientific understanding on this subject:

- Complex dynamical systems are composed of several interacting components.
- Many of the involved interactions are nonlinear and based on feedback interdependencies.
- The overall behavior of a complex system is an emergent property of those functional interactions.

From this point of view, biological systems are then complex systems.

Because of its scientific novelty, systems biology does not have a well-established and universally accepted definition. In fact, systems biology has two main different origins, each one characterized by its own idiosyncrasy:

1. Dynamical systems theory.
2. Computational biology and bioinformatics.

The first approach has in fact a quite old origin (Newtonian celestial mechanics, born at the end of the seventeenth century), while the second approach is more recent; it originated with the proliferation of electronic digital computer-based technology and the generation of large molecular data sets from high throughput genomic technologies during the second half of the twentieth century, e.g.:

- Trasncriptomics
- Proteomics
- Metabolomics

Let us briefly describe what it is understood by these two scientific fields:

**Dynamical systems theory:**    Also known as the mechanistic approach, considers that the behavior of any complex dynamical system is strictly determined by its past history, by the interactions of its constitutive components, and the modulation imposed by its surrounding environment. Moreover, for dynamical systems theory these interactions are regulated by fluxes of some sort of energy and information. Thus, a given physical dynamical system is understood basically as an energy this reference to energy can be confusing processor. The mechanistic approach is focused on the description and analysis of the dynamic behavior of the system resulting from the system's response to exogenous stimuli given initial conditions and the involved parameters.

**Computational biology:**    Describes biological systems as information processors, and privileges the study of the concerned system in terms of pattern recognition by decomposing the biological system in its constituents, mainly of biomolecular nature (e.g., transcripts, proteins or metabolites).

As far as computational biology is concerned, it is also common nowadays to consider it as the statistical detection of correlations between biological empirical data resulting from high throughput experimental explorations of the system under different conditions.

*Remark 2.11 (Dynamical Systems Theory as a Theoretical and Practical Choice)* In practice, any realistic exploration of biological systems needs systems-based tools developed by both dynamical systems theory and computational biology. As progress is made in both approaches, various opportunities to connect them are emerging. In this volume, to tackle human biomedical issues we choose a dynamical systems theory approach (as we previously pointed out). We believe that not sure this is a fact theoretical and practical tools developed around the formal study of dynamical systems constitute a theoretical framework well posed to uncover the basic principles that rule biological systems.

Let us now be more specific about the main methodological approach of our chosen scientific perspective.

## Bottom-Up and Top-Down Approaches

We are convinced that dynamical systems theory offers the right modeling framework that we require in order to satisfy our objectives. More specifically:

- the multistable nature of complex biological systems,
- the system's behavioral consequences in response to the presence of endogenous and exogenous stimuli,

are well-captured by this approach. This framework, i.e. the mechanistic perspective, is well posed to tackle the description of biological systems following a *bottom-up approach*. This modeling philosophy is defined as follows:

**Definition 2.10 (Bottom-Up Approach)**  This approach proceeds first through the characterization of the basic principles that rule the behavior of well-characterized basic components of the concerned biological system, and then through the piecing together of all these components in order to explore the behavior of the given system as a whole, or of modules of such system. Bottom-up approaches are grounded on well-curated experimental functional data for particular biomolecular interactions under particular conditions.

Even if we are interested in multistable systems (since we are concerned in this volume by cellular phenotypic plasticity), it is important to mention that not all biological systems are multistable. For instance, under some conditions related to the gain of an involved feedback-based interaction, some mitogen-activated protein kinase signaling cascades can display multistability or monostability (see [22]).

As far as the computational approach for systems biology is concerned, the basic methodology is known as the *top-down approach*. This modeling perspective obeys the following epistemological philosophy:

**Definition 2.11 (Top-Down Approach)**  Proceed first by piecing out all the constituents of the studied system, observing their behavior under different conditions, and then identifying via pattern recognition structural and functional connections or interacting modules, for example by exploring the correlation in the behavior (e.g., up or down-regulation) of different sets of components under contrasting experimental conditions or for different life-stages or tissues types.

*Remark 2.12 (Advantages of the Bottom-Up Perspective)* The selection of the guiding modeling approach is a matter of choice (in fact, it could even be a matter of personal preference). We prefer the bottom-up approach because it offers a mechanistic, rather than a descriptive, understanding of the phenomena involved. This mechanistic understanding is required when considering systems-level interventions on disrupted core regulatory networks.

Using the bottom-up approach and focusing on the characterization of basic functional components (as well as basic interactions), offers a potential opportunity for the conception of well-designed based-on systemic knowledge of functional interventions intended to understand and propose therapeutic practices that modulate the behavior of the concerned systems in the right direction. However, it is clear that eventually the bottom-up and the top-down approaches complement each together for a more complete understanding of biomedical systems.

Let us now add to the modeling perspective the networks-based approach issue.

### → Modular Decomposition and the Networks-Based Approach

When describing a given complex biological system, it is a common modeling strategy to decompose it in its well-identified constitutive elements (e.g., structural or functional modules). The system is then understood as a complex network of interacting modules. This lead us to the following:

**Definition 2.12 (Networks-Based Approach)** When a system is coded in terms of the specification of the system behavioral consequences of the interactions between the constitutive components (i.e., a well-defined collection of functional nodes), modeled as a network of interacting well-characterized components within modules and among interacting modules.

Since biological dynamical systems can be interpreted as networks, it is very tempting to study them in networks-based terms. We can identify at least two well-defined research fields turning around networks-based understanding of complex dynamical systems:

**Networks-based topological analysis:**   This approach privileges the study of the systems functionality in terms of specific patterns of interconnections between the constitutive elements. This methodology, which extensively applies graph-theory tools, looks for the identification of the specific role that individual nodes or motifs play on the overall dynamics of the network (see for instance [8] and the references therein).

**Networks-based process analysis:**   This second approach privileges the deep understanding of the specific dynamical processes carried out by the specific interactions between the constitutive nodes, and the dynamical consequences on the behavior of the whole system.

Having in mind the development of therapeutic methodologies, at this time networks-based topological analysis does not offer the required systems-level insight. That is the reason why in our case we are concerned by the second perspective. In practice, both approaches coexist.

In the previous chapter we introduced the notion of systems biology in colloquial terms. Now, and taking into account our previous discussion, we shall postulate the following definition:

**Definition 2.13 (Systems Biology)** Domain of the biological science that studies the interactions between the components of biological systems, and how these dynamical interactions in concert give rise to the structure, function, and behavior of the whole system, under the dynamical constraints imposed by its surroundings.

Since systems biology tackles the understanding of complex biological systems through the mechanistic perspective, we need the following:

**Definition 2.14 (Biological Mechanism)** A system of causally interacting well-defined parts and processes that produce one or more well-observed and well-characterized effects.

This definition implies that to name a biological mechanism is equivalent to explain its function. We name a biological mechanism when we have an explanation of what constitutes it.

*Remark 2.13 (Biological Systems as Machines)* In conceptual terms the previous definition has a descriptive purpose. We are not assuming at all that biological systems are just machines (in the common Newtonian mechanical sense for this term); we use the concept of biological mechanism as a tool to ease our understanding of complex biological systems. Hence, systems biology describes biological phenomena via the characterization of the set of well-established biological mechanisms that explain the specificity of the biological system.

We summarize in a graphic way the systems biology modeling framework that we put forward in Fig. 2.6.

When using dynamical systems, i.e., state-based mathematical models, as the chosen tool to describe biological systems, the involved biological mechanisms take the form of time-dependent equations. In the molecular case, if the concerned system consists of a network of interacting biological molecules belonging to several molecular species, the (causal) biological mechanisms underlying the behavior of the system usually take the form of a systems of chemical rate equations. In this case the state of the system could be the time-indexed list of the molecular concentrations of all the molecular species involved (if a continuous-time description formalism is chosen). In the case of a gene network, the involved biological mechanisms describe the set of causal interactions giving rise to the activation states of the genes being considered, and the state of the system could be the list of the activation states of such genes, possibly best described in discrete-time terms. In fact, the latter would be the case if the gene network is a transcriptional regulatory one, a kind of gene network that is central in our research approach, for the reasons that we explained in the previous chapter.

*Remark 2.14 (Holistic Understanding)* Even if the systems biology research methodology is based on the characterization of the biological mechanisms underlying the behavior (as well as the structure and the function) of biological systems, its working approach is not reductionist at all. In fact, for systems biology description is not good enough. Moreover, systems biology has not as its main purpose just to list the uncovered biological mechanisms. Systems biology looks for the holistic understanding of complex biological systems to uncover the dynamical consequences arising when all the explicit detected biological mechanisms are combined. This understanding needs to make the causal specific mechanisms, as well as the interplay between these biological mechanisms, in order to explain a complex biological system as a whole.

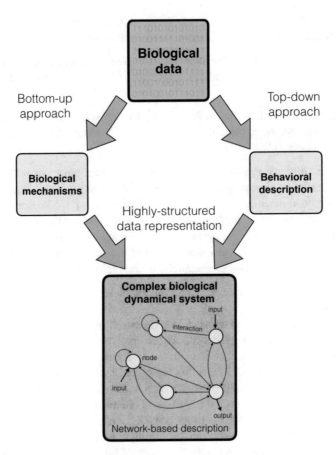

**Fig. 2.6** Modeling framework of medical systems biology. The medical systems biology modeling strategy goes from empirical biological data to the description and dynamical network-based analysis of the concerned complex biological system. The biological systems studied in this volume are composed mainly by biomolecules such as transcription factors, enzymes or extracellular ligands. Our chosen modeling framework follows a bottom-up approach, that begins with the characterization of meaningful biological mechanisms and that ends with the behavioral description of the systems, coded as a state-based dynamical system. We decided to follow this pathways to system's characterization, but it is not the only one. A top-down perspective (going from the available empirical data to the description of the system's behavior) also result in network-based system's characterization

## *Regulation as a Modeling Guiding Idea*

In general, it is very difficult to fully describe and understand complex biological systems. A more realistic goal is the description and understanding of specific classes of biological systems that regulate specific biological functions.

*Remark 2.15 (Functional Interplay Between Complexity and Modularity in Biological Dynamical Systems)*  Empirical evidence suggests that to deal with complexity, natural biomolecular systems are (broadly speaking) organized in modular terms, at both the structural and the functional levels (see for instance [92] and the references therein). Modularity makes then complexity functionally and computationally tractable. A fundamental consequence of modularity is regulation: the interaction between the constitutive modules needs coordination.

As an organizational principle modularity requires regulation. This task, the coordination of information exchange processes performed by the modules, is carried out by specific specialized regulatory networks. Such kind of systems can be defined as follows:

**Definition 2.15 (Regulatory Networks)**  A particular kind of biomolecular networks that regulate information exchange processes, underlying a specific functionality in the biomolecular systems under study.

Remember that in the previous chapter we colloquially discussed the relationship between the disruption of regulation and human diseases (and we exposed the fundamental role played by core regulatory networks, in the context of *developmental transcriptional regulatory networks*). We shall now insist in that topic, being more specific on certain modeling issues.

Strong evidence supports the epistemological claim that human disease comes as a result of disruption of the regulatory processes that sustain the organism at the biomolecular level is being accumulated. Transcriptional regulatory networks involve complex interactions of gene biomolecular products as well as biomolecules or signal transduction pathways that connect organisms with environmental cues. Indeed, it has been shown that diseases involve particular disrupted transcriptional core regulatory networks [267]. We schematically depict this statement in Fig. 2.7.

As exposed in the previous chapter, disruption is the dynamical consequence of a multi-factorial process perturbed by a set of risk factors. Let us now to argue how the understanding, and potentially the treatment, of disease can be tackled via a formal modeling systems-level perspective intended to uncover the mechanisms that rule this complex multi-factorial process.

### → Emergence of Disease

In modeling terms, attention must be focused on the interplay between:

1. Regulatory networks and
2. Genetic and (intrinsic and extrinsic) environmental perturbations

These disturbances alter network structure and function, disrupting regulation and eventually leading to disease.

This dynamical interplay can be understood in terms of modifications of the dynamics of the involved regulatory networks and their dynamical feedback with

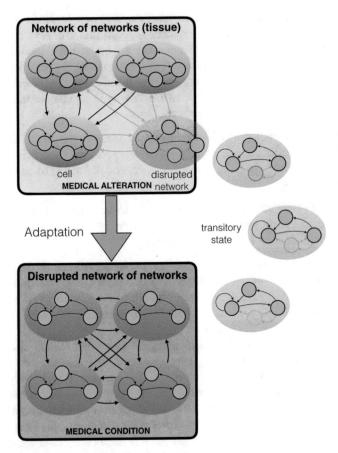

**Fig. 2.7** Disrupted regulation and disease. Thinking about chronic degenerative disease in terms of disrupted regulation requires to take into account how tissue integrity is lost. A coordinated community of cells (e.g., a tissue) can be seen in modeling terms as a coordinated network of interacting biomolecular networks. Each one of the involved networks sustains the functionality of the corresponding regulated cell. Disruption of the regulatory functionality of a given core transcriptional regulatory network might start a sequence of events giving rise to a consolidated new network of networks, supporting then the dynamics of a resulting community of cells adapted to the given unhealthy circumstance (constrained by a given specific environment at the micro and the macro levels). The dynamical process that goes from an alteration to a medical condition involves transitory states, where the phenotypic identity of is gradually modified until attaining a stable unhealthy state, through specific cellular state-space trajectories. Notice that the existence of transitory cell phenotypes reflects the plasticity of biomolecular networks

environmental conditions or factors (generally known as risk factors), and can be coded as state trajectories characterized by fragile initial states coding potential medical alterations and final states coding consolidated medical conditions. The transitory states that might be uncovered upon goal-oriented analysis of empirical evidence bridge the initial and the final states and result from the influence of environmental or risk factors (or both).

*Remark 2.16 (Topological Description of Human-Health)* In state-space-based terms, the addition of a distance quantifier between an unhealthy state at a given time and the corresponding desired healthy state makes the state-space a metric space (a particular class of what is known as a topological space). From a state-space perspective, it would be very advantageous to formulate human-health issues in terms of the quantification of the health-state as a well-defined topological distance (see Fig. 2.8). This would allow us to apply rigorous mathematical concepts in the conceptualization of a topological-based theory for medical systems biology. However, at this moment that ambition is out of the current state of both scientific research and medical practice, but current developments point in that direction.

One of the main advantages of a topological description of human-health is the application of principles of variational calculus in this context [258]. A given human-health-state trajectory can be then understood as the solution of a given variational problem, and a therapeutic intervention could then be seen as a specific variational optimization problem. This would require to code human-health in term of a formal optimization objective. However, this is easier said than done. In this volume we shall not go in that direction, but it is important to point out what other possibilities are opened by a state-space framework in medical systems biology.

It is time now to be more specific about a formal description of biological systems. For this we shall apply in what follows the systems-based perspective to the description of cellular functionalities.

## The Cell as a Dynamical System

Previously we defined what we consider a dynamical system in formal terms, now we apply that formal perspective and consider cells as dynamical systems. In other words, we assume that at any given time the measurable state of any given cell can be described by a set of time-dependent variables that satisfies the dynamical system definition that we put forward above.

We can describe the state of the cell at a given time by the finite collection of levels of expression or activation of the different molecular elements in it at a given time.

*Remark 2.17 (Molecular Diversity in the Mammalian Cell)* If we were to consider all the proteins in an average mammalian cell, we would be looking at around $2.7 \times 10^6$ proteins/$\mu m^3$ as the total number of protein molecules per cell volume [324]. Moreover, it has been estimated that there are 70,000 protein species in a single human cell [343]. Thus, a protein-based cell state descriptor would be a 70,000-dimension state vector. Studying the dynamics of such a large state vector is far from being realistic. Fortunately, such a complex system is under regulation. The cell

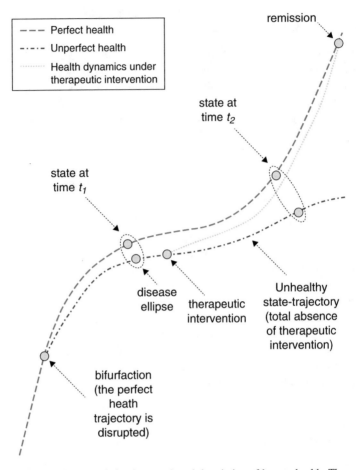

**Fig. 2.8** Conceptual framework for the state-based description of human health. The green line depicts the clinically defined good health trajectory, while the red line corresponds to a heath-state trajectory affected by a given disease condition. At a given time, say $t_1$, the length of the major axis of the disease ellipse represents the measured distance between perfect health and the affected health-state at that time. In absence of therapeutic intervention this disease will increase with time, as depicted with the disease ellipse at time $t_2$. Under medical treatment, a given therapeutic intervention is then applied to regulate the state-of-health in order to make it recover (if possible) good health. Ideally, remission is the goal of the therapeutic intervention

can be then seen through the dynamics of the involved regulatory networks. Thus, regulation offers us an efficient dimensional reduction procedure, which enables us to work with state vectors of analytically and computationally tractable dimensions.

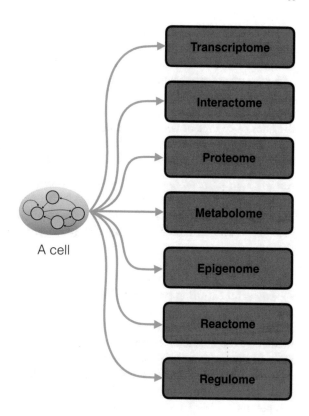

**Fig. 2.9** Omics information. Originally, the so-called Omics sciences dealt with the discovery and annotation of all the sequences in the entire genome of a given organism. Nowadays, Omics sciences are concerned by the construction of the whole set of cell biomolecular informational databases, as well as its goal-oriented analysis. These databases provide the raw data to develop network-based descriptions of cell biomolecular systems (e.g., transcriptional gene regulatory networks)

**Selection of the State Variable**

To pursue a state-space description of cell biomolecular network dynamics, we require to select a convenient state variable that is biologically relevant. Since transcriptional regulation is at the core of regulatory cellular networks, the state variable necessarily includes transcriptional factors and their interactions.

Nowadays it is easy to have access to high-dimensional cell biomolecular descriptive information that has been produced by high throughput approaches (see Fig. 2.9). This information includes data related with the dynamics of gene regulatory networks, which can be recovered through the application of specialized computer-based data mining methodologies (increasingly depending on deep learning and machine learning tools, see for instance [225, 229]) or by other approaches including well-curated functional data for sets of transcriptional regulators.

For practical reasons, the chosen state of the regulatory machinery of the cell has been commonly approximated by the levels of expression of the genes that encode for the transcriptional factors that are key for a particular cellular behavior or function. The levels of expression constitute a convenient, measurable quantity, from a systems dynamics point of view. We shall take such an approximation here.

For us each gene or transcript in the cell represents one time-dependent variable, and the genes interact with each other through regulatory mechanisms, therefore forming a functional network. If only the active and inactive states of the network's nodes are taken into account, a Boolean discrete-time description can be very useful to describe gene regulatory dynamics. In that case the descriptive state variable consists of a Boolean vector, which evolves in a finite state-space. However, if the rates of change of the involved biomolecular species (including gene products) are relevant, the state vector will consist of the solution of a particular system of ordinary differential equations. We shall continue this discussion later.

*Remark 2.18 (Extending the State-Space Perspective)* This gene-regulatory networks perspective can be naturally extended to encompass cells tissues or even organs or persons. To begin with, the state variable take into consideration the spatial location of chosen biomolecular species (e.g., gene products) or cells in specific tissues.

### $\rightarrow$ Powerful Intuition from the State-Space: Cell Phenotypic Identity

Before getting into more technical details, we will motivate showing the utility of modeling: the characterization of the state-space. The *state-space* is perhaps the main theoretical/conceptual tool that enables an intuitive yet rigorous investigation of some of the most fundamental questions in cell and developmental biology (as well as the systems biology view of human complex chronic degenerative diseases). Simply put:

> *The state-space is the abstract space where all the virtually possible cellular phenotypes reside.*

In reality we do not observe all the:

- possible expression profiles,
- cellular phenotypes,
- or morphologies.

Instead, we only observe a subset of robust and distinguishable cellular and tissue types, as well as reproducible patterns of pathological cellular and tissue conditions that allow clinical diagnosis.

Thus, our empirical observations of phenotypes in health and disease imply that some driving force should be maintaining cells within specific, restricted regions of the state-space.

As explained in the previous sections, in the mathematical theory of dynamical systems, such attracting regions of the state-space are called *attractors*.

*Remark 2.19 (Rectrictions and Cell Compatible States)* Irrespective of the initial phenotypic state, the restrictive behavior imposed by regulatory interactions within the gene regulatory network will push the cell away from incompatible states of gene activity and towards a specific set of attracting states defined by logically consistent gene activity configurations. For example, two mutually inhibitory regulators cannot be active and highly expressed at the same time, as imbalances in gene expression and stochasticity will always lead to one or the other direction—perhaps initiating a cascading regulatory transition and lineage choice or attractor switching.

The set of attractors (i.e., the stationary and stable gene configurations), underlie the possible observable cellular phenotypes or cell-types, which correspond to specific state-space neighborhoods that together form an emergent and organized structure that constraints all the possible and plausible patterns of cellular phenotypic transitions. Moreover, each attractor displays a characteristic context-dependent robustness-versus-plasticity balance.

> From this perspective, we know that disease emergence and progression is epitomized by either the appearance of pathological neighborhoods, or the occurrence of atypical, out-of-context transitions towards existing immature or normally unstable states due to the restructuring of the state-space.

The combination of intrinsic and extrinsic factors of both genetic and epigenetic or environmental origin induce such state transitions during normal development in healthy and ill individuals, or as part of rational therapeutic interventions.

The models discussed below provide tools intended to study and predict the structure of the state-space associated with specific experimentally grounded gene regulatory networks (that comprise the necessary and sufficient set of restrictions or interactions) under physiological and altered conditions.

## 2.5  Discrete Single-Cell Boolean Models

We start by discussing the simplest modeling frameworks used to operationalize the systems dynamics perspective of the cell: deterministic, discrete single-cell Boolean models.

In discrete-time dynamical models it is assumed that both the time and the state variables take discrete values. That is, it is assumed that at each time-step the state of gene activity can take only one of a discrete set of values or levels. Simplifying further, we can define the simplest discrete model by limiting the state variables, say $x_i(t)$ ($i = 1, 2, \ldots$, the total number of state variables), to take only binary values (i.e., $\{0, 1\}$), thus obtaining the widely used Boolean gene regulatory model (see for instance [240]). The mapping of cell state transitions functions then become:

$$x_i(t+1) = F_i(x_1(t), x_2(t), \ldots x_n(t)), \quad i = 1, \ldots, n \qquad (2.1)$$

where:

- The set of transition functions $F_i$ formalize logical propositions expressing the relationship between the genes that share regulatory interactions with the gene $i$.
- The state variables $x_i(t)$ can take the discrete values 1 or 0 indicating whether the gene $i$ is active or not at a certain time $t$, respectively.
- Positive integer $n$ denotes the number of nodes in the network (i.e., the dimension of the associated state-space).

Despite the high degree of simplicity and abstraction, the deterministic discrete-time description of Boolean networks provides a level of understanding useful for explaining actual observed developmental processes (e.g., cell differentiation and morphogenesis). Moreover, this modeling approach is able to predict outcomes under novel contexts in multiple experimental systems (See examples and reviews in [17, 19, 29, 49, 112, 318, 369]). The functional relationships coded in the transition functions $F_i$ can be readily obtained directly from experimental data (see for instance [19, 112, 478]).

Thus, Boolean gene regulatory networks provide a powerful framework to integrate diverse empirical data in the form of regulatory relationships and to computationally interrogate their systems-level behavioral consequences.

In what follows we first present the practical application and analysis of the Boolean gene regulatory approach, and then go on to expand this modeling tool to continuous dynamical models.


## *The Boolean Approach*

Boolean gene regulatory networks capture key qualitative aspects of developmental systems, while being simple and intuitively appealing [6, 80, 318]. In this subsection we provide a more detailed conceptual as well as practical presentation of discrete-time Boolean networks as a first step towards integrating complex biomedical systems.

In order to explain the formalism we:

1. Define a Boolean gene regulatory network model (selection of genes and identification of regulatory relationships).
2. Show how such model is computationally analyzed to recover its dynamical behavior.
3. Explain how the network attractor states are obtained and validated with experimental data.
4. Characterize the configurations of the state-space
   and:
5. Discuss how we validate the resulting models in terms of robustness and predictions of mutant phenotypes.

The mathematical basis of the models to be used can be found in [6, 80, 138, 235, 318].

### → Essential Components of Boolean Gene Regulatory Networks

A dynamical and deterministic discrete-time Boolean model representing a gene regulatory network has two essential components:

1. A list of genes hypothesized to summarize the properties of interest in a developmental system and a prediction of how those properties change over time. The list of chosen genes defines the state of the system $(x_i(t))$.
2. A set of updating rules (i.e., mapping functions $F_i$) specifying in terms of logical propositions how the activity of each gene changes over time as a function of the current activity of all the involved genes. Thus, the updating rules fix the system's transitions.

This approach can be applied to a broad class of biological systems (not only to developmental ones). Moreover, not only genes are involved in gene regulatory networks. Other regulatory players can also be playing an important role.

For Boolean networks, the length of the state vector corresponds to the number of nodes, and the total number of states that constitute the state-space (i.e., the time-dependent variables) equals 2 to the power of this length (see Fig. 2.10 for a sketchy illustrative academic example).

*Remark 2.20 (Network's Curation)*  Through extensive literature curation of functional experimental data, the genes and their regulatory rules can be proposed. The set of genes and updating rules should represent the current state of knowledge regarding what is known about the regulation of the developmental process or biomedical system of interest under study. It is quite common nowadays also to get the required data or to complement it from the databases generated by the Omics sciences (see Fig. 2.9).

Let us now specify the procedure giving rise to the construction of a deterministic discrete-time Boolean network model. The procedure is schematically summarized in Fig. 2.11.

### Defining the Set of Genes

In the first step in gene regulatory network modeling we need to define a set of genes (and the other involved biomolecular players) to be included in the network. Depending on the biomedical problem under consideration, a candidate regulatory module is postulated by proposing a set of molecular components known to behave as key regulators of the concerned process. The aim is to focus on a core regulatory module that has been associated to a particular developmental process, as explained in the previous chapter of this volume. We provide in this book two examples of

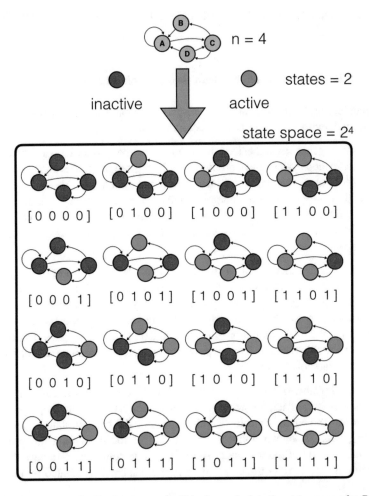

**Fig. 2.10** State-space of a Boolean network. This figure depicts the state space of a Boolean network constituted by four nodes, shown in the top of the figure. The nodes are represented by circles and the interactions between nodes are represented by arrows (i.e., the network is represented as a digraph in terms of graph theory). Each node has only two possible states: *active* (blue) and *inactive* (red). The depicted Boolean network has $2^4 = 16$ configurations. Each one of these configurations is represented via a binary row vector [$A$  $B$  $C$  $D$], and can be interpreted as a potential initial condition of the system

regulatory core modules that illustrate the usefulness of the Boolean approach (see the details in the next chapter):

1. A model of epithelial-to-mesenchymal transition underlying the emergence and progression of epithelial carcinoma.
2. A model of CD4+ T cell differentiation and plasticity during normal and hyperinsulinemic conditions.

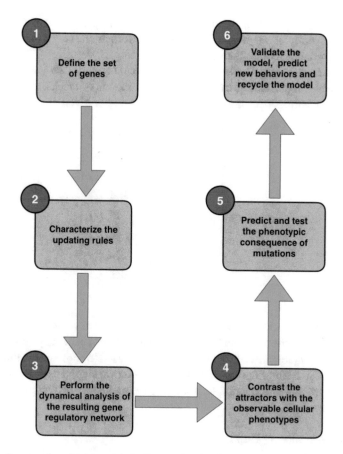

**Fig. 2.11** Construction of a deterministic discrete-time Boolean network model. This figure shows the six stages that constitute the general procedure that is followed when tackling the construction of a gene regulatory network grounded on empirical data. The procedure goes from the definition of the genes that define the state of the concerned network to the recycling of the resulting deterministic discrete-time Boolean network model

Consider, for example, the model of carcinogenesis. In this first model, extensive literature search and curation resulted in the discovery of a set of nine molecular key players regulating the molecular processes proposed to underlie carcinogenesis, namely (see the details in [316]):

**Cellular Senescence:**                  p53, p16.

**Cell-cycle:**                           TEL, E2F, Rb, cyclin.

**Epithelial cell differentiation:**      ESE-2.

**Mesenchymal cell differentiation:**     Snai2.

**Cellular inflammation:**                NF-$\kappa$B.

The nine genes involved in the core gene regulatory network were identified via a meticulous study of the available published information, under the guidance of experts on the subject of epithelial cancer.

In the corresponding Boolean gene regulatory network, we then define nine binary variables representing the activity configurations of the set of the identified nine genes. In symbolic terms, we have the binary vector given by:

$$\mathbf{x}(t) = \begin{bmatrix} x_1(t) \\ x_2(t) \\ x_3(t) \\ x_4(t) \\ x_5(t) \\ x_6(t) \\ x_7(t) \\ x_8(t) \\ x_9(t) \end{bmatrix} := \begin{bmatrix} \text{p53}(t) \\ \text{p16}(t) \\ \text{TEL}(t) \\ \text{E2F}(t) \\ \text{Rb}(t) \\ \text{cyclin}(t) \\ \text{ESE-2}(t) \\ \text{Snai2}(t) \\ \text{NF-}\kappa\text{B}(t) \end{bmatrix}$$

This set of variables $\mathbf{x}(t)$ represents the state of the cell at any given time. In the next chapter we present the biological and theoretical background of the problem, and the detailed and complete modeling approach that we used to formalize it that allowed us to tackle it in a simplified manner. In this section we shall only use this example to present a general introduction to Boolean gene regulatory network modeling.

**Defining the Updating Rules**

Now that the set of genes in the network has been defined (i.e., the set of key players involved in the regulation of carcinogenesis through the epithelial-to-mesenchymal transition dynamics), we need to propose an updating function for each of the genes.

The updating function will act as the dynamical mapping ($F_i$) connecting the present and future activity states of the corresponding $i$-gene. Altogether, these functions will update the state of the network each time they are applied in a synchronous manner (i.e., all the genes in the network make a simultaneous transition from the present state to the next state). Consider for example the gene telomerase TEL. A natural-language statement published in experimental studies might state that:

> Down-regulation of Snai2 and ESE2 induces increased expression of telomerase.

Suppose that another study reports that:

> Over-expression of Snai2 and ESE2 down-regulation induces increased expression of telomerase.

We can translate these two complementary statements into a simple logical proposition as follows:

$$\text{TEL}\,(t+1) = (\text{not Snai2 and not ESE2}) \text{ or } (\text{Snai2 and not ESE2}). \qquad (2.2)$$

> We can also write this logical proposition using the standard notation for logical operators, i.e.:
>
> $$\begin{aligned} \wedge &: \text{and} \\ \vee &: \text{or} \qquad\qquad (2.3) \\ \neg &: \text{not} \end{aligned}$$

Thus:

$$\text{TEL}\,(t+1) = (\neg\text{Snai2} \wedge \neg\text{ESE2}) \vee (\text{Snai2} \wedge \neg\text{ESE2})$$

The right-hand side of (2.2) represents a logical statement formalizing the nature of the influence of Snai2 and ESE2 over TEL. The given equation is the update function postulated to determine the dynamics of the activity of the gene telomerase (TEL). In words:

Telomerase will become or stay active if either both Snai2 and ESE2 are currently not active, or Snai2 but not ESE2 is currently active.

At first sight, it does not seem that such as simple rule will be able to generate any interesting dynamical behavior. However, when we consider that each time the activity of both Snai2 and ESE2 will itself be determined by two other updating functions that can potentially be much more complex and involve more regulatory molecules, each being regulated by an update function, then the coordinating power and nontrivial behavior of even simple Boolean gene regulatory networks becomes evident.

Once we extend the same type of reasoning to the simultaneous implementation of all updating rules corresponding to all the genes or nodes in the regulatory network, it is easy to imagine how the regulatory restrictions will ultimately determine small subspaces (i.e., well-characterized collections of network's states) consistent with all the constraints. This emergent restructuring of the state-space is what specifies the attractors and the corresponding observable cellular phenotypes.

*Remark 2.21 (Constraints and Robustness in Gene Regulatory Networks)* As an evolutionary product, the constraints that rule the interplay between the genes associated to the regulation of a specific biological function, coded by the updating rules in the case of Boolean gene regulatory networks, decompose the state-space and fix the cell-state trajectories. These constraints give rise to a level of determinism that explain robustness of gene regulatory networks, even in the presence of stochastic fluctuations of the involved biomolecules or some loss and gain of function mutations or weaker alterations of the logical rules.

Following the previous simple exercise of translating natural-language statements published in the relevant experimental literature to logical functions, we can define the complete set of updating rules and complete the definition of our Boolean gene regulatory network model. We must point out that the logical updating functions can readily be used as input for specialized existing computational tools to directly proceed with the analysis of their dynamical behavior. This is, in fact, one of the most important advantages of discrete-time Boolean networks as modeling tools of gene regulatory networks.

It is clear that the translation from natural-language statements to logical functions makes possible the development of computer-based automatic tools for the description of gene regulatory networks. This asks for the inclusion of Artificial Intelligence developments in the context of medical systems biology. In the spirit of what is illustrated in [284], both text mining and data mining of reported empirical evidence can be *automatically translated* to logical propositions, and then the collection of these propositions can be explored to shape regulatory networks.

### → Dynamical Analysis of Boolean Gene Regulatory Networks

A gene regulatory network is completely specified by the proposed set of nodes and their corresponding updating rules (that code the empirical evidence). Once any network is specified, it is possible to analyze its associated dynamical behavior. Consider as an illustrative example a network with eight genes (see Fig. 2.12a):

The state of all the genes specify a gene configuration of the complete network (network state $\mathbf{x}(t)$) at each time step $t$. In the Boolean case, a gene regulatory network with $n$ genes can only take state values from a finite state space of $2^n$ possible states—because each gene states can only take either of two values (0 or 1). Thus, the entire state space of a network with eight genes will have $2^8 = 256$ states. In Fig. 2.12 we schematically represent these states as circles in a quadrant, comprising the state-space. Each circle in the given state-space corresponds to a specific state vector or expression configuration with eight "0" or "1" activity values, one for each gene. The parallel implementation (i.e., all the genes in the network make a simultaneous transition) of the updating rules (i.e., the transition mapping) $F_i$ uniquely determines a specific future network state, starting from any initial state. For example, when only the gene $x_1$ of the network is active at the present state, application of the updating rules will produce a future state where the genes $x_1, x_2, x_3, x_7$ are now active, as shown in Fig. 2.12b. Since the state-space is finite and closed, if the updating rules are applied iteratively starting from a specific network state, it is possible for the network to reach either a stationary state that does not change after applying the updating rules (i.e., a fixed-point attractor) or a close set of states from where the updating rules map to themselves (i.e., a cyclic attractor). These stationary circumstances are the *attractors*. In the current illustrative example, once the network state where only the genes $x_1, x_3, x_7$ are active is reached, further implementation of the updating rules will not change

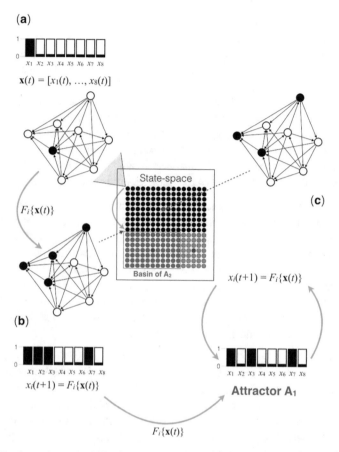

**Fig. 2.12** Boolean gene regulatory network dynamics. The figure schematically shows key concepts involved in the modeling of gene regulatory dynamics dynamics. (**a**) In the initial state of a network with $n = 8$ genes only the gene $x_1$ is active. This configuration constitutes a state that corresponds to a point in the state-space of the network, which is partitioned into three *basins of attraction*. (**b**) Application of the updating rules or mapping function results in a state change resulting in the activation of three more genes. The state change is reflected in a change of location in the state-space. (**c**) An additional updating of the network state results in a new state, where any additional updating results in the same state. Consequently, this last state is in fact an *attractor* state, and it is represented as a red point in the state-space. Each basin of attraction contains a different attractor state. Resulting from the action of the transition mapping the state-space is structured in a collection of state-regions, i.e., the basins of attraction associated to the set of attractors

the state anymore (see Fig. 2.12c). In other words, the attained attractor state is consistent with the regulatory constraints imposed by the regulatory logic coded in the updating rules.

Extending this logic, if the updating rules are applied exhaustively starting from every possible network state (i.e., the initial conditions), some subsets of states converging to the same attractor will emerge. The regions in the state space converging to the same attractor are called *basins of attraction*. Usually more than one of these basins of attraction emerge (i.e., the systems is multistable), thus as a result of the dynamical behavior, the state-space is effectively partitioned into attracting neighborhoods. In the current example (Fig. 2.12), the dynamics of the network converges to three fixed-point attractors (red circles), and, accordingly, the state-space is partitioned in three basins of attractions (black, blue, green); one for each one of the attractors. Remarkably, in a real model, by virtue of the nontrivial restrictions coded in the model specification, the uncovered attractors represent activity configurations that are consistent with the regulatory logic specified in the experimentally based updating functions—analogous to the way in which gene expression profiles observed in vivo maintain specific cellular phenotypes.

*Remark 2.22 (The Attractors Landscape)* As a consequence of the gene regulatory network dynamics, the state-space gets structured in a specific manner, which is reflected in the way the basins of attraction are organized. We will refer to such structure as the *attractor landscape*. In the case of development, the characterization of this landscape allows us to explore and study the patterns of cell-state transitions naturally emerging during developmental processes, as well as its potential manipulations through rational interventions. In the context of complex chronic degenerative diseases, attractors landscapes allow the construction of formal settings for the understanding of disease dynamics.

To summarize, in Fig. 2.12, we schematically show the key concepts involved in the modeling process of gene regulatory network dynamics. The implementation of the updating rules to a given network state will determine a specific network state for the next time step. The iterative implementation of the update functions to this latter state will eventually drive the network to reach an attractor state. The most basic characterization of the corresponding attractor landscape consists in obtaining all the attractors as well as their corresponding basins of attraction.

*Remark 2.23 (Open Access Computational Tools for the Modeling of Regulatory Networks)* A detailed mathematical presentation of dynamical systems theory for this type of model is out of the scope of this book. We direct the interested reader to existing superb references such as [429], and [157]. Fortunately, years of research and the popularity of gene regulatory models have led to the implementation of a multitude of open access modeling computational programs that can be used to analyze Boolean network models. Such tools already include efficient implementations of the necessary methods required to simulate gene regulatory network dynamics and to characterize the associated attractors landscape, as well as several approaches to conduct mutation and robustness analyses of network models. Some examples of useful tools for the dynamical analysis of Boolean networks are:

| **BoolNet**:   | [336] |
|----------------|-------|
| **ANTELOPE**:  | [25]  |
| **GINSIM**:    | [340] |
| **GNbox**:     | [97]  |
| **GNA**:       | [115] |
| **BioCham**:   | [71]  |

Various approaches to obtain the attractors and the technical details implied have been reviewed elsewhere (see for instance [163]). Moreover, we have previously published complete modeling protocols of gene regulatory networks, describing how to exploit existing computational tools to propose and analyze experimentally grounded networks (see for instance [30, 112, 466]).

**Contrast the Attractors with the Observable Cellular Phenotypes**

The main hypothesis of the modeling framework of gene regulatory networks is that:

> The experimentally grounded network constitutes a regulatory mechanism driving the specification of the experimentally observable cellular phenotypes.

In order to support such hypothesis one expects that the attractors uncovered through the dynamical analysis above will correspond exactly to the observable cellular phenotypes in terms of the configurations or profiles of gene activation states. The set of uncovered attractors thus defines predictions as to how cell-state configurations are expected to be, given the regulatory constraints integrated in the network. In order to test such model predictions, and thus evaluate the suitability of the model at hand, we have to empirically measure (or to retrieve from experimental literature) the actual, observable cellular phenotypes corresponding to the developmental system being modeled. For comparison, the observable phenotypes are represented in a format comparable to the binary state vectors recovered with the Boolean model. The latter will ultimately enable to bridge predictions and observations. If both predicted attractors and empirical phenotypes show close to perfect correspondence (due to some unavoidable circumstances, such as missing data as well as the variability of gene expression profiles, perfect correspondence is hard to attain), we postulate that the uncovered gene regulatory network includes a set of genes and regulatory interactions that naturally explain the observed phenotypes, suggesting an explanatory molecular systemic and dynamical mechanism. Note, however, that it is common practice in model building to go through multiple rounds of modification and testing, before uncovering the final, consistent gene regulatory network model.

**Prediction and Testing of the Phenotypic Consequence of Mutations**

Testing for consistency between predicted attractors and observed cellular pheno-types is a necessary first validation step. However, further computational validation analyses are often used to test whether the uncovered network constitutes a functionally relevant regulatory module. First, a naturally evolved biological mech-anism is expected to be robust against errors and random perturbations (see for instance [141] and the references therein). To test whether the uncovered model tolerates such perturbations, random errors can be simulated by introducing random modifications to one or several of the updating rules defining the given Boolean model. Subsequently, a comparison of the dynamical behavior of the modified and the original network will indicate if, and to what extent, the network dynamic behavior is conserved. A robust gene regulatory network is expected to show an unaltered dynamical behavior in response to a large number of tests.

*Remark 2.24 (Measuring Robustness)* In the context of discrete-time (and discrete-space) Boolean networks, as exposed in [29, 141], a standard measure of robustness quantifies the frequency at which the attractors uncovered with the original gene regulatory network exist in each perturbed network. Therefore, perfectly robust attractors of the original network are those that exist in all perturbed networks (resulting from errors and random perturbations affecting the original network), and totally fragile attractors of the original network are those that have a zero frequency of existence in the perturbed networks. Thus, if an uncovered regulatory network, described in Boolean terms, is such that all its attractors have a high frequency of existence in all perturbed networks, we say that the uncovered network is robust.

A gene regulatory network module that is consistent in terms of recovering observable cellular phenotypes, as well as displaying a robust behavior under random perturbations, provides an explanatory mechanism. The uncovered network can be further used to predict and test the cellular phenotypes that are expected to emerge under specific genetic perturbation. Boolean gene regulatory networks are particularly well suited for such predictive experiments, as it is straightforward to simulate single or combined loss- and gain-of-function mutations by simply fixing the value of a gene as 0 or 1, respectively. For the case of existing experiments analogous to the simulated mutations, direct comparison can be used as further validation of the predictive power of the uncovered model. In turn, a gene regulatory network model able to predict independent phenotypes as result of mutation strongly supports that an underlying biological mechanism has been uncovered.

**From Validation and Prediction to Model Recycling**

The ability to simulate novel mutations and to predict the corresponding phenotypes provides a valuable framework to prioritize candidates to be experimentally tested. Both robustness assessment and mutant simulation experiments further support the hypotheses implied in the gene regulatory network model, and are important steps in the modeling protocol. In addition to testing mutations, the uncovered gene regulatory networks can also be useful as a computational frameworks to test the impact of modulatory factors (or therapeutic interventions). Once the model has been validated, it becomes a useful framework to integrate newly available data; constituting a valuable building block for more complex models. New biologically meaningful data can be used to update the network based on modeling recycling (see for instance [167]). The idea is to propose updated models that are able to better explain observed experimental evidence. Such models also enable predictions that are likely to be more accurate in the context of the available data and the uncovered regulatory module.

*Remark 2.25 (Identified Gene Regulatory Network Rules as Hypothesis Testers)* There are cases in which the outlined protocol becomes more complicated. In some cases, different updating rules can be consistent with the data at hand. In fact, networks with different structures can produce the same combination of functionalities, being also able to produce similar attractor landscapes (see for instance [31]). In this context, the overall gene regulatory network module can be viewed as a "hypotheses tester" framework, with which the different options can be tested. It is always convenient to review and discuss such cases with experimental biologists (or biomedical professionals) who are experts in the molecular underpinnings of the cases under consideration. In many cases, additional arguments can be found to select among the possible alternative functions. The dimension of the space of alternative functions shrinks as the quality of the experimental evidence improves.

Given that in most systems under analysis it is common to find gaps or holes in the experimental data, alternative gene regulatory network models can be used to postulate alternative system-level hypotheses (in fact, an identified gene regulatory network module can be seen as a preliminary explanation of the phenomena under study). In any case, having a systems biology dynamic model is recommended to postulate alternative hypotheses and predictions; rather than just following an empirical trial-and-error approach because once the system under consideration goes beyond two components, the number of combinations and possibilities increases as double exponential (see for instance [138]).

## Networks Organization: Spatial Considerations and Coupling of Gene Regulatory Modules

The modeling framework described up to now does not incorporate any explicit spatial information. However, gene regulatory dynamics are intimately related to spatial constraints. In fact, in multi-cellular organisms cell communities possess high levels of spatial organization, and this robust organization involves physicochemical fields. Thus, mechanical forces emerge as a result of cell-to-cell interactions (like the ones generated by the action of *integrins* and *cadherins* that give rise to cell–cell adhesion forces), and the presence of mechanical constraints generates specific gene regulatory dynamics that condition the spatial organization of specialized cell communities (e.g., tissues). Moreover, the dynamics of the interplay between gene regulatory networks and mechanical constraints display feedback-based interactions with the whole organism's regulatory chemical fields (e.g., the human body's endocrine system regulates via the endocrine signaling processes the function of whole organs, which require the regulation of specific genes in each involved cell, see for instance [313]). From a modeling point of view, all this complexity seems to be overwhelming. However, once function-specific intracellular gene regulatory network modules are proposed, it is relatively easy to study the coupled dynamics of biological networks in explicit spatial and temporal domains. Hence, a meta-gene regulatory network model can be proposed to recover morphogenetic patterns. In the simplest case, a network (composed by a $N$ number of cell types or spatial locations in a lattice) of networks of dimension equal to the $N$ cells times the $M$ intracellular components can be proposed. In such meta-network each component is identified by the location or cell type in which it is found at a particular moment, and also by its gene identity. Cell-to-cell communication, due to active molecular transport biomolecules, and/or the action of chemical (e.g., diffusion) and physical fields, can be tackled via the specification of spatially dependent initial conditions ruling the dynamics of the coupled gene regulatory networks underlying the behavior of the cellular structure. Such spatial dynamical models are out of the scope of this volume, but we direct the interested reader to studies previously published by our group (see for an example, [29]). In such case, restrictions that depend both on intracellular networks, as well as on patterns of cell-to-cell communication can be explicitly considered to study both cell differentiation and spatial cellular patterning, i.e., where the cellular lattice is not static.

A next step would imply considering the following phenomena (among others):

- Cell proliferation;
- Multi-level modeling.
- The feedback from signaling pathways and microenvironment conditions or elicitors (which is addressed later).
- Cell-cycle or metabolic modules.

Since identified gene regulatory network modules do not work in isolation, modeling is necessarily confronted with the coupling issue. In the context of deterministic discrete Boolean gene regulatory networks, the coupling problem comes into play when more than one module has been identified (each of them consisting of a particular given set of logical rules), and these identified modules collaborate in a regulatory function that combines the regulatory functions fulfilled by each participant module (e.g., the gene regulatory module that coordinates the cell-cycle collaborates with the gene regulatory machinery that coordinates cardiogenesis in mammals [516]). Moreover, in a given cell, the whole set of gene regulatory network modules works in coordination to ensure the overall behavior of the cell. Couplings are unavoidable. Modularity, as a robustness-oriented

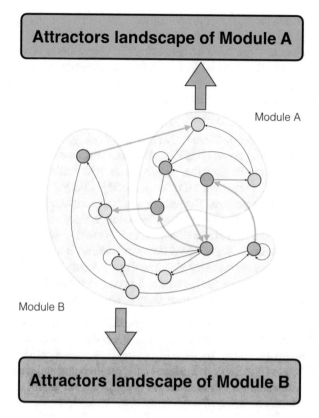

**Fig. 2.13** Coupling of gene regulatory network modules. This figure shows in a schematic way the logic behind the coupling of deterministic discrete-time Boolean gene regulatory network modules as discussed in [228]. Module A and Module B are considered to be autonomous systems when being isolated (i.e., they have not inputs and/or outputs), each giving rise to a particular attractor landscape. In isolation, Module A and Module B contain all the genes colored in orange and in green, but when interconnected a choice substantiated by meaningful biological empirical evidence of the location of the common genes must be carried out. This choice introduces inputs and outputs, as depicted in the figure, and gives rise to a global attractors landscape whose members can differ from the union of the members of the isolated attractors landscapes

organizational principle, is a complex phenomenon in the biological context (see for instance [362, 475]). Transcriptional networks are essentially modular, but this does not mean that the constitutive modules are just abstract dynamical systems exchanging information. In fact, it is quite common that interacting modules have shared nodes. Moreover, these shared nodes (i.e., specific genes) are not only bridges allowing interactions between modules but agents fully engaged in the fulfillment of the regulatory functions performed by the interacting modules (see Fig. 2.13). When applying deterministic discrete Boolean descriptions, this phenomenon guides coupling modeling procedures (like in [82]). The coupling issue is out of the scope of this volume. Fortunately, the available literature provides some formal methodologies to tackle this issue (while still requiring non-automatized knowledge-based techniques). A formal Boolean networks coupling procedure has been developed in [228], based on the developments presented in [82], conceived for the decomposition of large Boolean networks. The procedure is illustrated in [228] via the coupling of the epithelial-to-mesenchymal transition core gene regulatory network developed in [316], with the mammalian cell-cycle regulatory module developed in [144]. A coupling methodology based on logical regulatory graphs as well as transition graphs is included in GINsim [314].

*Remark 2.26 (Indirect Interactions)* Even if a set of gene regulatory network modules are not sharing common nodes, this does not mean that interactions are absent. This is due to the fact that the modules under consideration could be interacting by proxy. In other words, unknown bridges between the modules under consideration could be present. Remember that transcriptional networks interact with other types of networks (e.g., transcriptional regulatory networks interact with metabolic networks and signaling pathways). Indirect interactions make modeling a real challenge. In some circumstances a suspected indirect interaction between gene regulatory modules can me modeled via *virtual components*, which characterization can be an important challenge for the research agenda resulting from the analysis of the dynamical behavior of the gene regulatory modules under consideration.

Once a gene regulatory function has been described via a specific discrete-time Boolean gene regulatory network module grounded on empirical data, we can proceed to test hypothesis and to make biological meaningful predictions. For this, we take the discrete-time Boolean model as a departure point, and we perform some well-tested transformations of the model in order to extract useful information. In what follows we present some of these transformations.

## 2.6   Continuous Approximations to Discrete Single-Cell Models

As we previously discussed, the deterministic discrete-time (and discrete-space) Boolean gene regulatory networks have been very useful in the study of the complex logic of gene regulation involved in cell differentiation, as well as in the identification and explanation of cell-types as dynamical attractor states. However, for the study of more detailed dynamical behaviors (i.e., behaviors that do not only include like-switching dynamics), which are often determined by quantitative aspects of gene regulation, more complex mathematical models are needed. In particular, lumped parameter continuous models are very useful, as far as the study of transient dynamics are concerned. Therefore, as a modeling recycling idea it is handy to have a way of transforming existing discrete-time Boolean gene regulatory networks into continuous dynamical models (then coded by differential equations). This because continuous-time and continuous-space descriptions allow the consideration of transient dynamics involved in regulatory dynamics, as well as the inclusion of parametric information. Although multiple approaches considering different levels of detail are often used in systems biology to model via differential equations, we consider here the most general and simple methods that enable obtaining continuous approximations of the discrete Boolean models of gene regulatory networks (see for instance [30]). More detailed, mechanistic continuous models will be introduced in the following sections.

## *From Switching Dynamics to Smooth Saturated Behaviors*

The direct mapping of discrete-time Boolean dynamics to continuous-time dynamical models allows extending the bottom-up gene regulatory network modeling approach at multiple levels of spatial and temporal resolution. Importantly, the obtained continuous models (taking the form of systems of *ordinary differential equations*) enable exploring different types of questions and generating novel quantitative predictions and comparing results with those derived with the Boolean framework. More specifically, in the clinical context, continuous-time models are useful when considering parameter-dependent transient dynamics, for instance:

- The interplay between androgen dynamics and the dynamics of prostate tumor [220].
- The quantitative description of type 2 diabetes and obesity [11].
- Systems-level dynamics dependent on environmental constraints (e.g., evolution of cancer cells under the influence of the concentrations of nutrients and therapeutic drugs [61]).

In what follows we summarize how to approximate a deterministic discrete Boolean gene regulatory network into a continuous dynamical system. The analysis of such differential equation models will be the focus of the next sections.

In the general case, in the absence of exogenous stimuli (a very idealistic situation) the dynamics of deterministic and continuous models of gene regulatory networks is given by a system of autonomous ordinary differential equations. In such case, the time evolution of the cell-state:

$$\mathbf{x}(t) := \{x_1(t), x_2(t), \ldots, x_n(t)\}, \tag{2.4}$$

is modeled by a system of ordinary differential equations of the form:

$$\frac{dx_i(t)}{dt} = F_i(x_1, x_2, \ldots, x_n, \mathbf{p}), \quad x_i = 1, 2, \ldots, n, \tag{2.5}$$

where:

- $x_i(t)$ denotes the concentration of the $i$-th product resulting from a corresponding gene expression process.
- $dx_i(t)/dt$ corresponds to the rate of change of the concentration of $x_i(t)$.

A system's dynamics defined by such a system of ordinary differential equations is a special form of the general transition map linking present and future states, in this case, where both time increments and state variables take continuous values (recall the formal definition of dynamical systems discussed in the previous chapter).

Importantly, this definition also introduces a parameter vector **p**. This vector is included to quantify the regulatory relations among genes. This allows us, for example, to introduce temporal gene hierarchies, as well as analyze qualitative changes to the state-space through bifurcations [112], as we will see in the next sections. Thus, the first goal is to simply map directly into a deterministic continuous-time dynamical model the set of updating rules describing a discrete-time Boolean gene regulatory network. To this end, consider decomposing the functions $F_i$ in (2.5) into a system of ordinary differential equations of the form:

$$\frac{dx_i(t)}{dt} = \Theta[f_i(x_1, x_2, \ldots, x_n)] - k_i x_i, \tag{2.6}$$

where:

- $f_i$ denotes the logical regulatory function that characterizes the dynamics of the $i$-th gene.
- $k_i$ represents the expression decay rate of the $i$-th gene of the given gene regulatory network.
- $\Theta[f_i]$ denotes a transformation that maps the discrete-time Boolean dynamics coded by the logical function $f_i$ into a continuous-time function.

As we have shown before [30, 112], a simple and useful way to transform the switching functions that shape the Boolean dynamics is by applying the following mapping rules (see (2.3)):

$$\left\{\begin{array}{l} x_i(t) \wedge x_j(t) \rightarrow x_i(t)x_j(t), \\ x_i(t) \vee x_j(t) \rightarrow x_i(t) + x_j(t) - x_i(t)x_j(t), \\ \quad\quad \neg x_i(t) \rightarrow 1 - x_i(t). \end{array}\right\} \tag{2.7}$$

In other words, by substituting the logical operations in the Boolean update function with the arithmetical operations following the rules above, we can effectively map the discrete-time Boolean space to the continuous-time state realm [30].

Thus, instead of having rules dictating whether the binary activity state of one gene will change (or remain the same), the transformation allows us to quantify a regulatory continuous-time input for that gene. In order to process this input into an activity output of the gene, we consider an input-response gene function that displays a smooth saturation-like behavior as is customary in modeling chemical reactions. Formally, the input associated to a $i$-th gene is included in the form [30]:

$$\Theta[f_i(x_1, x_2, \ldots, x_n)] = \frac{1}{1 + \exp\left(-b\left(f_i\left(x_1, x_2, \ldots, x_n\right) - \epsilon\right)\right)},\qquad(2.8)$$

where:

- $\epsilon$ is a threshold level (usually $\epsilon = 1/2$), and
- $b$ is the input saturation rate.

Note that for $b >> 1$, the input function displays step-like behavior, getting close to the binary on/off behavior of the Boolean case. In simple words, this transformation smooths a switch-like behavior in which the regulatory input of a gene produces an all-or-none (i.e., either active or inactive) response into a $s$-type function where the output is of quantitative and continuous character.

Thanks to this simple transformation, for any deterministic discrete-time Boolean gene regulatory model, we can obtain a corresponding finite set of ordinary differential equations that respects the regulatory restrictions included and validated in the Boolean case, and that can be subjected to all the well-developed toolkits for modeling and analyzing differential equations. This allows various quantitative approaches to explore, for instance:

- Robustness properties of the associated dynamical attractors.
- Structural fragilities of the underlying network of interactions.

More specifically, for each gene the regulatory term that involves the corresponding gene decay rate acts as a negative regulation of that node with a strength given by the value of the associated decay rate.

*Remark 2.27 (The Hill Function and Glass Transformation)* There are also others methodologies to transform discrete-time Boolean models to continuous-time descriptions. For instance, the Hill function and the Glass transformation (see [169, 488]) are well-known alternatives. From a modeling perspective, a selection of a qualitative-oriented methodology that involves a low-dimensional parameter space should be preferred over an alternative that requires an important number of parameters to be tuned in a precise manner.

## Parametric Dependencies of the Discrete to the Continuous Transformation

When transforming a given deterministic discrete-time (and discrete-space) Boolean model to a continuous-time (and continuous-state) model (coded by ordinary nonlinear differential equations), we add to the resulting description a required set of parameters. Sometimes these required parameters do not need to be realistic in order to be useful (this does not mean at all that the parameters values are just arbitrary!). As far as gene regulatory networks are concerned, the expression decay rates of the involved genes condition in a differentiated manner the level of involvement of the genes in transient dynamics (and then in the fate of the concerned cell). Indeed, actual gene decay rates are not constant but context-dependent variables (see for instance [243]). In order to extract significant information concerning the transient dynamics, the transformed system can be analyzed through computer-based simulations, and these involve explorations in the parameter space (see [112]). To pursue this, the analysis of the transformed system can be performed specifying a range of values for the parameters.

One of the most important advantages of discrete-time Boolean models, as qualitative descriptive tools, lies in the fact that they constitute an abstraction of rates of chemical reaction. Moreover, empirical evidence shows that actual transcriptional regulatory networks are functionally close to discrete-time Boolean networks *almost independently from the values of the involved parameters* (see for instance [289, 374]). Moreover, all-or-none dynamics in transcriptional regulation are supported by the intervention of epigenetic regulatory mechanisms [300].

Before continuing with the analysis of ordinary differential equations in the next sections, we discuss, in what follows, an extension to discrete Boolean models in order to consider the ever present stochasticity of biomolecular systems and their functional cell-fate developmental consequences.

We shall show how to exploit a discrete-time Boolean gene regulatory model as a computational tool to explore transient dynamics due to stochastic influences.

## 2.7   Stochastic Cell Population Models and Epigenetic Landscapes

As discussed previously, stochastic models consider uncertainty in the dynamical outcomes by considering random variables as descriptors of the system's behavior. In other words, the same set of regulatory restrictions and initial conditions do not always produce the same state change. Rather, the change is influenced by variables whose outcome is uncertain at any given time. This uncertainty can represent unknown processes potentially affecting changes in cell behavior or sampling errors due to a limited number of molecules involved in the regulatory interactions being modeled. That is, stochasticity can be seen as an operational approximation to deal with the unavoidable incomplete information about the system (i.e., stochasticity can be considered as a modeling artifact in order to consider unmodeled dynamics). However, uncertainty can also constitute an intrinsic property of the system under study. For example, different cells in a population produce different mRNA molecules at different times, and these changes occur in discrete bursts that produce variability or biological noise. Irrespective of the actual nature of uncertainty, it is important to have effective ways to model it and to account for its potential functional consequences (particularly on transcriptional regulatory dynamics). As we will see, uncertainty of any source can be naturally included in the Boolean gene regulatory model, enabling then the study of cell states and state transitions in terms of probabilities.

## *Modeling Uncertainty Through Stochasticity*

The standard analysis of the deterministic models presented up to this point mostly focuses on recovering the network's attractors and characterizing their local properties (e.g., the characterization of the corresponding basins of attractions). As a consequence of this exhaustive local characterization of the properties of attractors, the geometry of an underlying attractors landscape also emerges. However, in such deterministic setting, under fixed values for the related control parameters and a given fixed initial regulatory network state, the cell always reaches a specific attractor, and remains in such attractor, if the network is not disturbed. This is because any chosen initial state necessarily belongs to the basin of attraction of an attractor, and the whole state-space is the union of all the attractors, and their corresponding basins of attraction. However, a given developmental path requires the concerned network to display a corresponding state motion shaped by a trajectory between a set of attractors. Then, the commonly observed plasticity of development, where cell-state transitions recurrently occur in different directions, is more naturally captured by considering *uncertainty*. Noise in resonance with the deterministic kinetic interacting functions can lead to the gradual movement of the state variable (i.e., the cell's phenotype) between attractors, resulting in the emergence of transient dynamics, corresponding to developmental paths from one attractor to another one in a time-ordered pattern. Taking into account uncertainty, we are interested here in stochastic dynamical models. Stochastic models enable the study of potential transition events among cell states, even under fixed parameters and an initial state. Due to the introduced uncertainty, the cell system can reach and surpass the boundaries of a state associated with a given phenotypic state that is defined by or corresponds to a particular attractor.

*Remark 2.28 (Uncovering Transitory Dynamics via Stochastic Explorations)* The implementation of stochastic models, in conjunction with the nonlinear constraints of the concerned regulatory network, enable the study of signal-independent transitions among attractors. In other words, stochastic analysis provides tools to uncover the dynamical consequences of a system's uncertainty and to address whether recovered time-ordered patterns correspond to the generic developmental paths or spontaneous temporal patterns that are observed in vivo.

### Epigenetic Landscapes

Stochasticity also enables departing from an exclusive focus on the local properties of the network's attractors to instead characterize the global dynamical consequences of the underlying attractors landscape. The characterization of the attractors landscape provides a natural formalization of the classical metaphoric model of the epigenetic landscape first proposed by Conrad Hal Waddington in [473]. As we have highlighted in previous sections, in addition to generating the cellular phenotypic states (attractors), the constraints imposed by the underlying gene regulatory network also partition the state-space, determining a landscape of cell states with different levels of stability. The formalization of the epigenetic landscape in this context is conceptually simple:

> *The number, depth, width, and relative position of the attractor's basins of attraction epitomize in the realm of dynamical systems the hills and valleys of the metaphorical epigenetic landscape.*

Thus, for our purposes, the characterization of the attractors landscape corresponds, in practical terms, to the characterization of the epigenetic landscape.

Under uncertainty, we can estimate "how easy" it is to transit from one attractor to another by means of stochastic dynamics. We can further generalize and estimate the relative stability of the different attractors, thus establishing a natural hierarchy of transitions as a natural consequence of the geometry of the attractors landscape, and a fundamental result of regulatory constraints. Simply put, in addition to the determination of the attractors themselves, the most likely hierarchy of time-ordered transitions among attractors also emerges from the regulatory constraints imposed by the interactions that link the genes in a particular regulatory network module under analysis.

In order to exploit the global character of the analysis of stochastic systems, and to characterize the dynamics of transitions across attracting neighborhoods in the state-space, in what follows we present modeling extensions to the discrete-time Boolean approach with the aim of characterizing the epigenetic landscape associated with the gene regulatory network. Remember that the epigenetic landscape is the multidimensional and multistable space that emerges from a nonlinear regulatory network. The basins of attraction and the attractors lie in such space. The latter restricts the possible transition pathways from one attractor and basin of attraction to another one.

## *Modeling Uncertainty in Discrete-Time Boolean Gene Regulatory Networks*

A given discrete-time Boolean gene regulatory network can be naturally extended into a discrete stochastic model by introducing uncertainty using the so-called stochasticity in nodes (SIN) model [112]. In this proposed model, a constant probability of error $\xi$ is introduced into the deterministic Boolean functions as follows:

$$\left\{ \begin{array}{l} P_{x_i(t+1)}[F_i(\mathbf{x}_{reg_i}(t))] = 1 - \xi \\ P_{x_i(t+1)}[1 - F_i(\mathbf{x}_{reg_i}(t))] = \xi, \end{array} \right\}, \tag{2.9}$$

where:

- $P_{x_i(t)}$ represents the probability of state at time $t$ and
- $\mathbf{x}_{reg_i}$ denotes the set of regulators of gene $i$.

This model considers that the state of a gene $x_i(t+1)$ is represented by a random variable, and the probability that its activity value is determined or not by the associated logical function $F_i(\mathbf{x}_{reg_i}(t))$ is $1 - \xi$ or $\xi$, respectively. Simply put, we assume that due to uncertainty stemming from either incomplete knowledge or intrinsic molecular processes, the regulatory rules will not apply unequivocally all the time, but instead there is a probability $\xi$ of observing an unexpected result (i.e., the state of a gene changes in contrast to what its associated logical rule specifies).

*Remark 2.29 (Stochasticity and Flexibility of the Regulatory Logic)* Stochasticity in this setting effectively provides flexibility to the regulatory logic by enabling a more plastic mapping between states. Note, however, that the regulatory constraints imposed in the regulatory function $F_i$ still channel the dynamics, so that the stochastic force is not independent but interacts with the underlying nonlinear dynamics. In the simplest model, the probability $\xi$ is a fixed parameter affecting independently each gene in the network.

We shall now illustrate how the proposed uncertainty model allows the study of the concerned network's transient dynamics.

## *Estimating Transition Probabilities of Attractors*

By simulating a stochastic one-step transition multiple times, according to the proposed stochastic model, and starting from each of all the possible states in the system, we can empirically estimate the probability of transition from one given attractor to another. The frequency of times the states belonging to the basin of the attractor $i$ are mapped into a state within the basin of the attractor $j$ under uncertainty constitutes a right approximation of the probability of such transition. Utilizing this intuitive simulation scheme, we can operationalize the stochastic Boolean gene regulatory model as a simple discrete-time Markov chain (MC) model (see [462] for a detailed exploration of stochastic processes), by defining an attractor transition probability matrix $\Pi$ with components:

$$\pi_{ij} = P(A_{t+1} = j | A_t = i),$$

representing the probability that an attractor $j$ is reached from an attractor $i$. These components are estimated from the empirical transition probability resulting from the performed simulations.

In this discrete stochastic dynamics, changes from one attractor to another are represented as a sequence of random variables $\{C_t : t \in \mathbb{N}\}$, where $C_T$ takes as values the characterized different attractors, and the elements $\pi_{ij}$ represent the inter-attractor transition probabilities. The resulting matrix $\Pi$ is then the (one-step) transition probability matrix that operationalizes for each attractor the likelihood of going to or coming from any of the other attractors. Note that the simplest model assumes a homogeneous MC, that is, the characterized probabilities do not depend on time.

The statistical behavior of the stochastic dynamics is represented at the ensemble level by the attractor occupation probability distribution $P(C_t = j)$, which represents the probability that the cell population is in attractor $j$ at a given time $t$, and it is denoted by the row-vector $\mathbf{u}(t)$. Considering the previously estimated inter-attractor transition probabilities $\Pi$, the attractor distribution of the cell population temporally evolves according to the dynamical equation:

$$\mathbf{u}(t + 1) = \mathbf{u}(t) \times \Pi.$$

This dynamical mapping equation enables simulating the dynamics of the cell population by simple iteration. However, by taking $\mathbf{u}(0)$ as the initial distribution of the MC, the equation reads $\mathbf{u}(1) = \mathbf{u}(0) \times \Pi$, and by linking the occupation probabilities iteratively we get:

$$\mathbf{u}(t) = \mathbf{u}(0) \times \Pi^t.$$

Thus, the occupation probability distribution at time $t$ can also be obtained directly by matrix exponentiation. In either case, the initial distribution $\mathbf{u}(0)$ represents the initial cell state distribution of the cell population—that is, whether all the cells belong to the same state of differentiation (same attractor) or are heterogeneous (fractions with different attractors).

Having obtained the temporal evolution of the occupation probability distribution $\mathbf{u}(t)$ given an initial distribution $\mathbf{u}(0)$, following [18], it is assumed that the most likely time for an attractor to be reached is when the probability of reaching that particular attractor is maximal. Therefore, the temporal sequence in which attractors are attained is obtained by determining the sequence in which their maximum probabilities are reached, i.e., max $\mathbf{u}(t)$. This sequence characterizes, in temporal terms, how the population of cells is organized (from the population defined by the initial attractor) throughout the developmental trajectory. In the context of biological development, this is an important consequence of the proposed analysis. Therefore, this methodology provides the characterization of most-probable developmental paths, offering then test-hypothesis opportunities. It constitutes a stochastic exploration of the epigenetic landscape and it was first proposed and applied to study floral organ specification time-patterning in [18].

## Developmental Paths: Emergent Hierarchies of Cell State Transitions

In addition to the calculation of the most probable temporal cell-fate pattern, a discrete stochastic gene regulatory network model also enables the calculation of the shortest and fastest pathways of cell-fate transitions, as well as possible restrictions of some cell-fate transitions. Both the likely and restrictive developmental pathways emerge from gene regulatory network constraints reflected in the topography of the associated epigenetic landscape.

One way to characterize developmental paths is by calculating the feasibility of sequential attractor transitions. A natural metric to quantify the latter is by statistically estimating how long it takes to transit from one attractor to another, under the constraints of the underlying gene regulatory network. The mean first passage time (MFPT) (i.e., the time taken for a random walker to reach a specified target), is a natural metric to quantify the feasibility of each pair of possible transitions. The MFPT can be estimated numerically by simulating a large number of samples of paths, simulated as a finite Markov chain process using the transition probabilities among attractors $\Pi$. More formally, the MFPT from one attractor $i$ to another $j$ corresponds to the average value of the number of steps taken to visit attractor $j$ for the first time, given that the entire probability mass was initially localized at attractor $i$. By estimating the MFPT for all the pairs of possible transitions, we can know whether from an initial attractor state it is more likely to observe first a transition to another attractor $i$ and then to attractor $j$, relative to all the other transitions (including reversal transitions). This directionality of developmental paths can be simply quantified as a probability flow, as proposed in [519]. Using the computed MFPT values, a net transition rate between attractors $i$ and $j$ can be defined as:

$$d_{ij} = 1/\text{MFPT}_{ij} - 1/\text{MFPT}_{ji}. \tag{2.10}$$

This quantity effectively measures the feasibility by which the system transits from one attractor to another as a net probability flow.

*Remark 2.30 (Directionality of Sequential Attractor Transitions)* The sign of the net transition rates in (2.10) naturally imposes directionality to the transitions. A directional path through the epigenetic landscape emerges naturally as the sequential order of states for which all involved net transition rates have a positive sign.

Interestingly, the differentiation paths might constitute robust emergent phenotypes, potentially biased by associated genetic backgrounds and/or environmental inputs to the network. For example, we recently applied the epigenetic landscape formalism to analyze the observed in vivo developmental phenotypes of flowers for several mutant lines. By quantifying differentiation paths based on stochastic gene regulatory network dynamics, we identified a preferential transition to a undifferentiated phenotypic state, providing a novel systems explanation for apparently incompatible evidence [369]. The same framework can be used to analyze the development of pathological versus normal conditions in humans.

*Remark 2.31 (Stochasticity Model Implementation)* We are aware that the implementation of uncertainty in Boolean gene regulatory networks might seem challenging, as it, requires tailored simulation schemes. In order to naturally extend the conventional analysis of Boolean gene regulatory networks, in [110] we recently implemented a set of tools for simulation and downstream analysis of stochastic Boolean dynamics that enable straightforward implementation of the models conceptually described here.

The proposed model of uncertainty, intended to estimate the attractor transition probabilities (as well as the most probable developmental temporal trajectories), shows the power of discrete-time Boolean networks as descriptors of actual gene regulatory networks grounded on empirical evidence.


## Reshaping of the Epigenetic Landscape

So far we have discussed a model of the epigenetic landscape based on the relative stability properties of the attractors recovered though the analysis of gene regulatory network dynamics. The practical implementation of such a model involves four steps:

1. Specification of an experimentally grounded dynamical model representing a gene regulatory network.
2. Characterization of the attractor landscape through deterministic dynamical modeling.
3. Computational estimation of cell state transition probabilities.
4. Analysis of the prevailing paths of sequential cell state transitions.

This modeling framework has been shown to be useful for the study of robust developmental processes under normal and altered conditions [18, 316, 369]. In both cases, the dynamics develops over the complex structure of a fixed epigenetic landscape.

An alternative view has been advocated by others (as is the case in [146, 379]) and explored by us in the context of deterministic but continuous dynamical models (as discussed in [112]). In this view, a dynamically changing (as opposed to fixed) epigenetic landscape is proposed as a potentially more accurate description of certain developmental processes. Structural changes in the epigenetic landscape might occur at time-scales similar to those in which the developmental process unfolds. For instance:

- environmental factors;
- intercellular communication;
- mechano-elastic forces,

might bias net transition rates between cell states, reflecting shape changes in the epigenetic landscape.

A simple framework to start exploring these possible scenarios is to first transform a well-characterized Boolean gene regulatory network into a continuous model (see previous section), and explore the impact of parameters into the structure of the underlying epigenetic landscape. For example, gene-wise numerical bifurcation analysis using the characteristic decay rate of each gene as a control parameter, and each attractor as initial state, can uncover gene perturbation able to qualitatively modify the structure of the epigenetic landscape [112]. As we will discuss in the next section, a qualitative change of the state-space as a function of a parameter is formally known as a bifurcation. Characterizing such dynamical changes in the state-space might help to answer questions regarding the effect of signaling in normal development, or rational interventions over the system. In such cases, the direct modeling of gene regulatory dynamics through a continuous-time formalism provides a more natural framework.

It is time now to address the issue concerning mechanistic descriptions of regulatory interactions using the continuous modeling framework.

## 2.8   Mechanistic Continuous Models

We have previously discussed the versatility of discrete-time Boolean models as modeling tools that describe the essential dynamics of gene regulatory networks. We have shown how the attractor landscape of a discrete-time Boolean gene regulatory network can describe the phenotypic plasticity of a cell. Moreover, we have shown how to include in the discrete-time (and discrete-state) description a simple uncertainty model via a stochastic extension, and how to extract from this extension important information concerning transient dynamics: the estimation of attractor transition probabilities and the most probable developmental temporal trajectories (i.e., *finite attractors sequences resulting from stochastic disturbances on the network*). As far as the understanding of high-level dynamical properties of gene regulatory networks are concerned, the discrete-time (and discrete-space) Boolean formalisms are then very successful. Still, there are some important questions turning around regulatory dynamics that cannot be answered by the proposed discrete Boolean formalism. In particular, *such modeling approach cannot explain how and when a specific logic emerges from specific biomolecular interactions*. Fortunately, this kind of questions can be tackled via a direct mechanistic continuous-time and continuous-state modeling approach. In what follows we shall consider this issue.

## *When to Use Continuous-Time and Continuous-State Models?*

Phenotypic transitions are the basis of the onset and progression of chronic degenerative diseases. As we have stated, discrete Boolean networks are powerful models
that can effectively describe multistable systems' dynamical behaviors, helping
then to explain how multiple cellular phenotypes, and the transitions among them,
emerge from the interplay between biomolecules, *even in the absence of precise
kinetic information*. This makes discrete Boolean network modeling framework a
very useful conceptual tool to tackle the study of the interplay between phenotypic
plasticity and the dynamics of disease dynamics. As we have previously discussed,
the key assumptions of Boolean models are:

1. The regulation between molecules can be described by combinations of logical
   operators.
2. The state variables are binary (i.e., can have two values, 0 or 1).
3. The dynamics of the system occur in discrete-time intervals.

Indeed, experimental evidence suggests that many biological regulatory networks
show such a discrete-time, discrete-state dynamical behavior, and that the regulatory
interactions can be described by logical functions. But:

- Where does such a discrete-state and discrete-time regulatory logic come from?
- What are the specific underlying biomolecular regulatory mechanisms?
- What are the effects of changes in kinetic constants on the dynamical behaviors?
- How are continuous changes in concentrations of regulatory molecules digitally
  encoded, by binary switching of the responsive biomolecular components?

The importance of these questions in some clinical contexts cannot be denied.
We shall show in what follows that continuous modeling allows to tackle these
questions, which cannot be answered with a Boolean framework. Moreover, we
shall illustrate when it might be useful to tackle medical systems biology research
using kinetic nonlinear ordinary differential equations.

### → A Priori Versus A Posteriori Logical Operators

Consider that we want to represent that the expression of gene $x$ can be induced by
the transcription factors $y$ and $z$. If we chose to describe this simple regulatory
interaction between $x$, $y$, and $z$ with a Boolean function, we could write the
model as:

$$x(t + \delta t) = y(t) \wedge z(t), \tag{2.11}$$

which means that only when both $y$ and $z$ are present, $x$ is transcribed.

Another alternative could be that gene $x$ can be transcribed when, $y$, $z$, or both
are present. If that was the case, then we would write the model as:

**(a)**

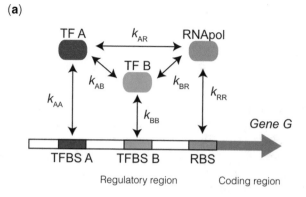

Protein-DNA association network

**(b)**

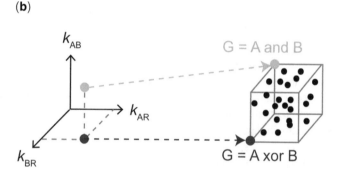

Mapping kinetic parameters to transcriptional responses

**Fig. 2.14** A posteriori logical operators. (**a**) The levels of transcription of gene $g$ depend on the association of the RNApolimerase (RNApol), and transcription factors (TF A and TF B), to their binding sites (TFBS A and TFBS B, respectively) on the regulatory region of gene $g$. The interaction strength is described by the kinetic parameters $k_A$, $k_{BB}$, and $k_{RR}$. The binding of these proteins to the DNA might also depend on the formation of heteromeric protein complexes (described by the kinetic parameters $k_{AR}$, $k_{AB}$, and $k_{BR}$). (**b**) Different kinetic constants quantifying the contribution of specific biochemical mechanisms shape the dynamical response of the network. Only in extreme cases these functions can be approximated by logic gates ($G(t) = A(t - \delta t) \wedge B(t - \delta t)$, $G(t) = A(t - \delta t) \vee B(t - \delta t)$, or $G(t) = A(t - \delta t)$)

$$x(t + \delta t) = y(t) \vee z(t). \tag{2.12}$$

But, hold on. How in the first place can such an *and* (multiplicative) or *or* (additive) function be achieved? In other words, which *biochemical interactions* between genes $y$, $z$, and the promoter of gene $x$ lead to an additive, and which to a multiplicative output function? Before solving our simple Boolean model, let us take a step back and consider some of the actual possible biochemical mechanisms by which the transcription factors $y$ and $z$ can regulate the expression of $z$. Some of the options are (Fig. 2.14):

**Option 1:**    The promoter of gene $x$ has Transcription Factor Binding Sites (TFBS) for both $y$ and $z$. So, if $y$ enters the cell nucleus it can:

- Bind to its corresponding TFBS,
- Recruit RNA polymerase (RNApol) to the promoter and hence
- Drive the transcription of gene $x$.

Similarly, when $z$ enters the cell nucleus it can bind to its corresponding TFBS and induce the expression of gene $x$. When both of them enter, each of them can independently bind to its TFBS and contribute to the expression of $x$.

**Option 2:**    Although the promoter of gene $x$ has TFBS for both $y$ and $z$, $y$ can enter the nucleus and bind to its TFBS and recruit RNApol **only** when it is forming a heterodimer with $z$.

**Option 3:**    The TFBS for $y$ and $z$ are overlapping; only one TF can bind at a time.

As can be seen, each option implies a specific regulatory network topology. Some questions arise when taking them into account:

- Which of these mechanisms could result in an *and*-type of input function, and which in an *or*-like function? In other words, how to map biochemical mechanisms and promoter architectures to transcriptional response functions?
- Which of these mechanisms can be approximated by a Boolean framework, where abrupt changes, from 0 to 1 (or from 1 to 0), occur in response to (continuous) changes in input concentrations? In other words, when are intermediate concentrations of inputs and outputs negligible?
- Which minimal concentrations of TF $y$ and $z$ are needed for triggering a *sharp* switch-like transitions of the system's output (expression of gene $g$ in this example)?
- When are these significant levels of expression achieved, and does this time lag $\delta t$ depend on the kinetic rates of binding and unbinding of the TFs with TFBS and RNApol?

These and other questions addressing the finer details of how the logical operators (that form the building blocks of Boolean models) *emerge* from the underlying biochemical interactions can, per construction, not be answered by Boolean formalisms. But fortunately for us, these can be explored by *mechanistic* continuous-time and continuous-state mathematical representations of the different possible biochemical interactions regulating gene expression. Such a mathematical framework is given by (kinetic nonlinear) ordinary differential equations (ODE's), which allow the explicit representation of biochemical interactions between biomolecules (using for this the well-known Law of Mass Action), in continuous-time and continuous-state.

Indeed, ODE models have been used extensively to explore how different network topologies describing differential binding between TFs and TFBS map to discrete logic gates (see for example [305, 404, 410]). In other words, these models show under which conditions Boolean operators emerge a posteriori from biochemical interactions (see Fig. 2.14).

### → Dynamical Consequences of Distinguishing Between Inhibitory Mechanisms

Consider that we want to model the experimental observation that the levels (concentrations) of protein $P$ at time $t_1$ are inversely correlated with the levels of protein $Q$ at time $t_2 > t_1$. If we chose a Boolean formalism, then we could simply write:

$$Q(t_2) = \neg P(t_1), \ (t_2 = t_1 + \delta t) \qquad (2.13)$$

But, hold on. *How* is protein $P$ leading to a reduced concentration of $Q$? At least two possible mechanisms could underlie this relation:

**Mechanism 1:**   $P$ is a transcriptional repressor, and thus inhibits the transcription of the gene encoding $Q$.

**Mechanism 2:**   $P$ is a protease; it induces the degradation of the protein $Q$.

Although modeled with the same Boolean function, these two mechanisms differ strongly in terms of the dynamic response of the output ($Q$) to changes (increases) in the input, $P$. In the first case (inhibition of transcription), reductions of $P$ at time $t_1$ are reflected in an increased $Q$ only after enough time, say $\delta A$, has passed to allow the de novo transcription of $P$ (see Fig. 2.15, left). In fact, it is this delay in the transcriptional response that justifies the use of discrete time jumps when modeling gene regulatory networks. When using differential equations, such time delay can be explicitly represented with a delay term (see for instance [328]).

In turn, as shown in Fig. 2.15, right, if the inhibitory mechanism of $Q$ by $P$ is by induction of degradation, the time between administration of $P$ and decline of $Q$, say $\delta B$ is much shorter (i.e., $\delta B \ll \delta A$), asymptotically decreasing with the increase in the degradation rate. Thus, the use of an explicit delay term might be less justified. Therefore, care should be taken when coding different biochemical mechanisms with the same regulatory logic.

### → Discrete Filtering of Continuous Signals: Multistability in Discrete Versus Continuous Frameworks

Phenotypic plasticity (i.e., the one-to-many map of genetic-to-functional configurations), is a fundamental feature of cells (or the whole organism). It allows:

1. The adaptation of cells/organisms to different environmental conditions [303, 360, 465, 491].
2. The emergence of complex, multicellular structures [36, 330].
3. Ontogenetic changes [18, 86, 318].
4. Tissue regeneration [205, 348].

However, this phenotypic plasticity also underlies the pathological transformation of affected tissues that are characteristic of the onset and progression of chronic

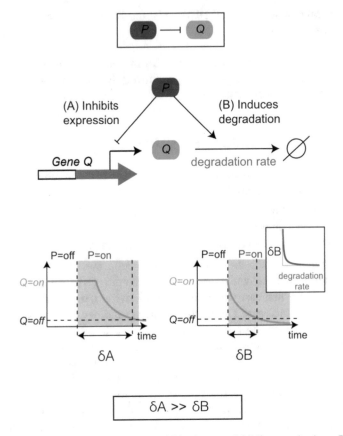

**Fig. 2.15** Dynamical consequences of distinguishing between inhibitory mechanisms. Regardless of the underlying biochemical mechanism, the inhibition of $Q$ by $P$ can be represented by the Boolean function $P(t) = \neg Q(t - 1)$. However, when timing matters, distinguishing between the mechanisms of inhibition might be important. (**a**) Shows the mechanisms of inhibition due to inhibition-by-repression. (**b**) Shows the mechanisms of inhibition due to induction-of-degradation. While inhibition-by-repression takes a long time due to the delay, say $\delta A$, that is intrinsic to the de novo transcription of a gene, when inhibition occurs by induction-of-degradation, increases in the repressor $P$ are reflected almost immediately (after a time delay $\delta B$) in a decrease in $Q$, i.e., $\delta B \ll \delta A$

degenerative disease, as we discussed in the previous chapter (see for instance [84, 217, 233, 316, 425]).

In dynamical terms, as has been previously discussed, different phenotypes can be interpreted as different stable steady-states, or dynamical attractors, of an underlying nonlinear regulatory network controlling cell-type-specific gene expression (see for instance [217, 233, 330, 465]). To obtain these attractors, these nonlinear regulatory networks must be translated into dynamical equations, from which the steady-state behavior can be characterized. Because of their simplicity (both in terms of their formulation and computational implementations), Boolean models offer an attractive framework to explore how multistability emerges from regulatory networks. Further, as discussed extensively above, Boolean formalisms can be used to characterize the topology of the basins of attraction (epigenetic landscape analysis, as discussed in [18, 112]). By doing this analysis for nominal versus genetically perturbed versions of the regulatory network, the effects of structural alterations on the network ("mutations") in terms of the existence, stability, and accessibility (topology) of attractors can be systematically evaluated. With a Boolean framework, it is also possible to characterize the minimal strength of transient environmental forcing necessary to drive the state of the system from one attractor to another one (see ahead and [85, 303]). This type of analysis is very useful to evaluate the mechanisms of phenotypic transitions underlying ontogenetic, physiological, and plasticity, as well as pathological processes.

*Remark 2.32 (Going Beyond Boolean Descriptions: The Bifurcation Analysis)* Given that Boolean network models are qualitative representations of regulatory *structures*, it is not possible with such an approach to study the effects of parametric changes on the system's behavior. *Bifurcation analysis* can, however, be important to understand how quantitative changes in the network can lead to qualitative transitions (i.e., bifurcations), how these bifurcation structures are shaped by other parameters and how the susceptibility to change from one attractor to the other in response to environmental perturbations (forcing) is affected by this bifurcation parameter [22].

In order to perform a bifurcation analysis, we require *quantitative* descriptions of the regulatory networks, such as nonlinear ordinary differential equations, for which extensive mathematical and computational theory for the bifurcation analysis has been developed (see for instance [22, 183, 224, 256, 429]). Such a qualitative bifurcation analysis can lead to:

- the detection of early warning signals that predict an imminent bifurcation [40, 84];
- the development of patient-specific, personalized biomarkers and treatment options [149, 195, 450];
- the design of optimal pharmacological intervention strategies that effectively (de)stabilize specific phenotypic attractors using the minimal amount and duration of the treatment [88, 122].

These are some of the advantages that bifurcation analysis provides to the agenda of medical system biology.

Let us now discuss how models coded as systems on nonlinear ordinary differential equations allow the detection of system's dynamical properties that depend on the (quantitative) magnitude and duration of environmental stimuli.

→ **Modulating Transient Responses**

Negative feedback has long been associated to the ability of a system to maintain a nominal, operational, or homeostatic state, even in the presence of perturbations in the form of environmental or genetic fluctuations [426, 493, 505]. Classical examples of such systems are:

- The synthetic Tet-repressor system in *Escherichia coli* [47, 346].
- The pheromone response pathway controlled by Fus3 in *Saccharomyces cerevisiae* [93, 385].
- The BMP4 signaling pathway modulated by the pseudo-receptor BAMBI in *Xenopus laevis* [365].
- The signaling networks underlying bacterial chemotaxis [14, 91].

In mathematical terms, such an adaptive response to perturbations is reflected in the ability of the output response variable ($R$ in Fig. 2.16) to return to a nominal steady-state ($R_{ss}$ in Fig. 2.16) after a *transient* deviation triggered by a change in the input conditions ($S$ in Fig. 2.16). For example, the perfect adaptation network discussed in [459], given by the activation of a regulator $X$ and a response $R$ by the input (stimulus) $S$, and a inhibition of $R$ by $X$, shows that, upon a step-like increase in $S$, $R$ transiently rises from $R(0) = R_{ss}$ to $R_{max}$ and returns then to a nominal steady-state $R_{ss}$.

*Remark 2.33 (Quantitative Assessment of the Dependency of $R_{max}$ and Time-to-$R_{ss}$ on Increasing $S$ Concentrations)* Both Boolean and differential equations models of this simple network are able to capture this long-term adaptive response. However, only the quantitative, continuous time-and-state framework of nonlinear ordinary differential equations captures the dependency of $R_{max}$ and time-to-$R_{ss}$ on increasing $S$ concentrations. Since values of $R$ and time-to-$R_{ss}$ might affect the onset of their parts of the system (for example, leading to the transition from one attractor to another if the perturbation is strong enough), implementing a continuous, tunable version of this $S$-to-$R_{max}$ and time-to-$R_{ss}$ map could have important functional consequences [3].

It is time to discuss now how the continuous description of regulatory networks can be very useful to uncover the dynamical consequences of the relative strength of the different regulatory interactions.

→ **Quantitative Modulations Shape Qualitative Transitions**

Phenotypic commitment of cells occurs through nonlinear signal processing of microenvironmental conditions by regulatory networks. As we have seen, the resulting input–output response depends of the topology of the network. Some of the quantitative features of this input–output relation between microenvironment and phenotypic response are in turn shaped by the relative strength of the different regulatory interactions, which can be affected by genetic (mutations, polymorphisms)

**(a)**

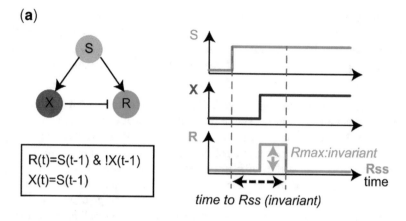

$$R(t)=S(t-1) \ \& \ !X(t-1)$$
$$X(t)=S(t-1)$$

**(b)**

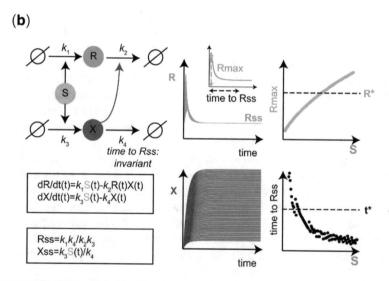

$$dR/dt(t)=k_1S(t)-k_2R(t)X(t)$$
$$dX/dt(t)=k_3S(t)-k_4X(t)$$

$$Rss=k_1k_4/k_2k_3$$
$$Xss=k_3S(t)/k_4$$

**Fig. 2.16** Modeling perfect adaptation networks with Boolean algebra (**a**) or differential equations (**b**). The perfect adaptation network discussed in [459], given by the activation of a regulator $X$ and a response $R$ by the input (stimulus) $S$, and a inhibition of $R$ by $X$, shows that, upon a step-like increase in $S$, $R$ transiently rises from $R(0) = R_{ss}$ to $R_{max}$ and returns then to a nominal steady-state $R_{ss}$. Although both Boolean (**a**) and kinetic differential equations (**b**) models of this network capture this adaptive behavior, only the qualitative, continuous time-and-state framework of ordinary differential equations captures the dependency of $R_{max}$ and time-to-$R_{ss}$ on increasing $S$ concentrations. Since the values of $R_{max}$ and time-to-$R_{ss}$ might affect the onset of other parts of the system, a continuous, tunable version of this $S$-to $R_{max}$ and time-to-$R_{ss}$ map could have important functional consequences

or microenvironmental (i.e., cell context) variations. These perturbations result in quantitative variations in network topology, which can have important functional consequences. The most common dynamic responses to an input, and how they are affected by quantitative variations elicited by genetic or contextual perturbations, are as follows.

1. In a linear cascade, the output dynamics follows the input, with a delay caused by intermediate states in the signal processing network. Network variations can result in differences in the time-to-maximum and time-to-relaxation. Despite having the same network topology, context-dependent variations in this regulatory network can lead to different phenotypic outcomes. For example, it has been shown in [125] that the memory of previous exposures to $\alpha$-pheromone in *Saccharomyces cereviceae* can be inherited from one generation to the next by differentially affecting the degradation rate of a signaling molecule that determines the response to this pheromone (phenotype decision: to mate or not to mate).
2. Negative feedback leads to a controlled output even under persisting input. Amplitude and duration of the transient response as well as the difference of output steady-states in the absence versus presence of persistent input (error) can be shaped by different input strengths and parametric variations (this is explained in detail in Fig. 2.16). Understanding how parametric perturbations shape the transient pulses of the output can be important, for example, when only high-amplitude or long-duration transient responses lead to a further phenotypic progression (as discussed in [122]), or when the fidelity of a signal determines the phenotypic response [93, 385].
3. Positive feedback (with cooperativity) can stably fix a phenotype even after the removal of the input driving the phenotypic decision. In such a multistable behavior, different reaction strengths determine the minimum time and duration required to fix the new phenotype (time-to-sepparatrix) (see Fig. 2.17). This might be important to characterize the patient-specific sensitivity of deleterious phenotypic transitions in response to environmental perturbations [122], or to find optimal, personalized treatment strategies that reverse such a pathological progression (as discussed in [88, 122]).
4. An odd number inverter can generate sustained oscillations [37], characterized by an amplitude and frequency (1/period) that can be modulated by quantitative variations in the reaction network. These variations in the amplitude and frequency of the response can have important functional consequences, in terms of the type of phenotypic response to the input condition. For example, it has been reported that the *input-specific response of some "master transcriptional regulators"*, that is, TF that can be activated by many different upstream signaling molecules, activated in turn by different inputs, and that have a myriad of potential transcriptional targets, can be explained by the capacity of such master transcriptional regulators to filter specific oscillation amplitudes or frequencies. Examples of master regulators for which such a specificity has been experimentally and theoretically shown include:

- NF$\kappa$B [381, 443].
- p53 [44, 45, 380].
- The MAPK pathway elements [395].

→ **Recap: When to Use Continuous Models for the Description of Gene Regulatory Networks**

As we have established before, *qualitative* models of biological networks are extremely powerful tools that allow the exploration of how different phenotypic features emerge from these networks. Many interesting and clinically relevant questions can be addressed with such an approach, including:

- Phenotypic plasticity (existence of multiple attractors; including multiple-state attractors, i.e. oscillators);
- The effects of structural perturbations affecting the behavior of different nodes of the network (mutations, environmental inputs);
- The time-ordering of the progression of pathological and ontogenetic cell states (characterization of the epigenetic landscape).

However powerful, this framework does not allow the examination of important biological phenomena such as:

1. The biochemical mechanisms underlying the regulatory network architecture.
2. The effects of parametric variations on the systems behavior (where these re- "bifurcation parameters" can be interpreted as patient-specificity [149], cell-specific context [70], or a treatment regime that should be optimized).
3. The effects of different kinetic rates on the system's behavior.
4. The role of intermediate concentrations of the involved regulatory molecules.

Reiterating, a mechanistic, quantitative, continuous-time-and-continuous-state framework that can be used to address these kinds of questions are ordinary differential equation models constructed on the basis of mass action kinetics. Since these models allow direct representations of biochemical mechanisms (building blocks of such models are the rates-of-change, which are direct representations of the most basic biochemical mechanisms: degradation, polymerization, post-translational modifications, etc.). Importantly, the regulatory logic does not have to be assumed a priori. As these models depend on parameters, the effects of quantitative variations on the network topology behavior can be assessed by many different mathematical and computational tools, including:

- Bifurcation analysis.
- Sensitivity analysis.
- Parameter optimization.

It is time now to introduce the main modeling tools to construct, analyze, and validate such mechanistic nonlinear ordinary differential equations-based models.

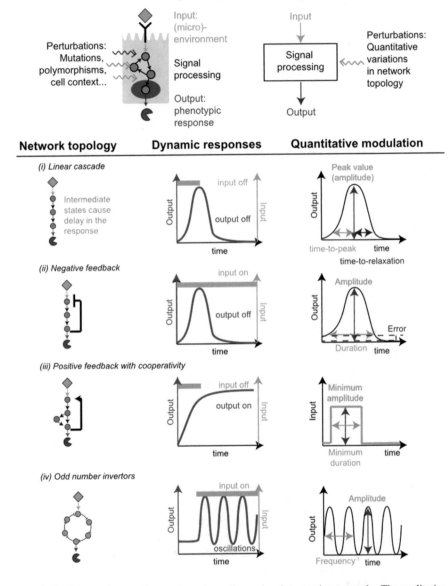

**Fig. 2.17** Phenotypic commitment through nonlinear signal processing networks. The qualitative features of the input–output relation between microenvironment and phenotypic response is determined by the topology of the corresponding signal processing network, and is shaped by the relative strength of the different regulatory interactions, which can be affected by genetic (mutations, polymorphisms) or microenvironmental (cell context) variations. These perturbations result in quantitative variations in network topology, which can have important functional consequences. (i) In a linear cascade, the output follows the input with a delay caused by

## Building and Analyzing Mechanistic Nonlinear Ordinary Differential Equations Models from Scratch

We introduce here continuous-time and continuous-state mechanistic models, namely kinetic models based on nonlinear ordinary differential equations. Using a simple illustrative example, we explain the pipeline for the formulation, parameter calibration, and analysis of a system of nonlinear ordinary differential equations in the context of gene regulatory dynamics modeling. The pipeline is formed by the following steps:

1. Visual representation of the biological system, using visual conventions that facilitate the direct translation of the network to a corresponding system of equations.
2. Mathematical representation of the biological system: translation of the network to a system of ordinary differential equations.
3. Simplification of the mathematical model (i.e., identification of conservation equations).
4. Identification of:

   - initial conditions;
   - parameters;
   - experimental data,
   - qualitative behaviors to be reproduced by the model.

5. Finding the equilibrium behavior of the system (i.e., steady-state analysis).
6. Dynamical simulation of the system of ordinary differential equations (integration).
7. Parameter optimization (i.e., seeking the best agreement between the mathematical model and the experimental data, using minimization algorithms).
8. Model analysis, i.e., assessing the robustness/plasticity of the model behavior in response to parametric variations, via:

   - Perturbation analysis;
   - Parameter sensitivity analysis;
   - Bifurcation analysis.

---

**Fig. 2.17** (continued) intermediate states in the signal processing network. Network variations can result in differences in the time-to-maximum and time-to-relaxation. (ii) Negative feedback leads to a controlled output even under persisting input. Amplitude and duration of the transient response as well as the difference of output steady-states in the absence versus presence of persistent input (error) can be shaped by different input strengths and parametric variations. (iii) Positive feedback (with cooperativity) can stably fix a phenotype even after the removal of the input driving the phenotypic decision. In such a multistable behavior, different reaction strengths determine the minimum time and duration required to fix the new phenotype (time-to-sepparatrix) (iv) An odd number invertor can generate sustained oscillations, characterized specific amplitudes and frequencies, can be modulated by quantitative variations in the reaction network

In what follows, each of these points is briefly described and exemplified. It is noteworthy that there are some software tools for systems biology purposes (both open-source and commercial) that help to perform most of the steps along this pipeline. For example:

Copasi [315]:                                    http://copasi.org
SBiology Toolbox for Matlab [405]:   http://www.sbtoolbox.org/

### → Step 1: Visual Representations of Reaction Networks

It is a good practice to start the construction of a mathematical model with an intuitive visual representation of the system. In the case of mechanistic models coded as nonlinear differential equations, it is very useful to follow the visual conventions given by the Systems Biology Markup Language (SBML) community [265] since there is an (almost) one-to-one relation between the graphical and the mathematical representation of the biological system. In fact, software such as Cell Designer (http://www.celldesigner.org) automatically creates ordinary differential models from a mechanistic user-defined graphical representation of the biological system.

As shown in Fig. 2.18a, the basic building blocks of a system of ordinary differential equations are:

- *Constants*. The value $c$ of the constants (*per definition*) does not change, and thus there is no need to write a dynamic equation for it (they are represented in the mathematical model by parameters).
- *Inputs*. The values of the inputs $u(t)$ can change, but this change is independent of the system's dynamics—it can be controlled by external conditions (for example, by the experimenter). The dynamics of the input $u(t)$ is represented in the mathematical model by algebraic equations.
- *Variables*. The values of the variables $y_i(t)$ change dynamically, as functions of the reactions. The system of equations describing the coupled dynamical changes of all the variables considered has the form:

$$\frac{dy_i}{dt} = F_i(y_1, \ldots, y_n, t), \quad i = 1, \ldots, n.$$

- *Outputs*. Correspond to the subset of the system's variables that represent the experimentally observed features of the system that one wants to reproduce with the model (for example, levels of target gene expression). While mathematically, it is not required to distinguish between the outputs and the other variables, it might be useful to make that distinction clear in the reaction network.
- *Reactions*. The reactions $r_j$ encode the biochemical mechanisms or processes regulating the system's variables. They are the building blocks for the functions describing the dynamical changes of the variables; the functions $F_i$ are linear combinations of the reactions, i.e.:

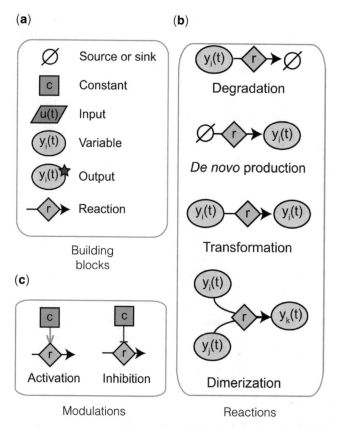

**Fig. 2.18** Basic building blocks to represent mechanistic reaction networks based on kinetic interactions between biomolecules. This figure shows the graphical elements that ease the visual representation of reaction networks. (**a**) Shows the basic building blocks, (**b**) includes the basic network motives, and (**c**) shows the modulations. A specific reaction network can be visually coded via these blocks, and then be automatically translated to the corresponding system of nonlinear differential equations, following the Systems Biology Markup Language (SBML) community conventions [265]

$$F_i\,(y_1, \ldots, y_n, t) = \sum_{j=1}^{m} r_j, \quad i = 1, \ldots, n. \tag{2.14}$$

- *Sources or sinks.* Are used to represent substrates or products of reactions that are not explicitly considered in the model. For example, the substrate of a de novo transcription of a gene are nucleic acids seldom coded as a variable and therefore representing a source or degradation of a protein, which produces aminoacids that are commonly neglected, thus represented by a sink.

Each reaction describes the transformation of a (set of) precursor(s) *a* to a (set of) product(s) *b*. Thus, while the concentrations of *a* decrease when this reaction occurs, $b$ increases with this reaction. A typical example of such a reaction is the conversion of a substrate into a product. Reactions can also be modulated by another molecule of the system and this explicitly depends on them. Positive modulators increase the rate of the reaction, and negative modulators decrease it. Modulators are not consumed or produced by the reaction they affect. A typical example of such modulator is an enzyme. Modulators $(x)$ can be a constant, an input, or a variable. The effect of $x$ on the reaction can also be additive (i.e., the reaction occurs even in the absence of $x$, like an *or* Boolean function), or multiplicative (the reaction occurs only in the presence/absence of $x$, like in a logic *and* Boolean function).

We can at this level proceed to the Step 2 of our pipeline.

### → Step 2: Mathematical Representation of the System

To construct a mathematical model that represents the reaction network, each of the regulatory interactions must be translated into a mathematical expression, particularly, a rate. Collectively, these rates form a system of differential equations that describe the inter-dependent dynamics of the different components of the reaction network. Translating a reaction network into a system of ordinary differential equations is a standard methodology in systems biology that has been discussed widely, for example in [100, 335, 429, 435]. The basic principle used to construct the individual rates of the system is the *Law of Mass Action*. It assumes that rate of change in the concentration of species $X$ is proportional to the concentration of precursors $X_{pre}$, the effectors $E_{reaction}$, and the kinetic rate constants (denoted by $k_{reaction}$ in each case), thus:

- A production reaction of $X$ is represented by the term:

$$\frac{dX(t)}{dt} = X_{pre} k_{prod} E_{prod}.$$

Since the production of $X$ from $X_{pre}$ also consumes $X_{pre}$, this reaction also negatively affects $X_{pre}$, i.e.:

$$\frac{dX_{pre}(t)}{dt} = -X_{pre} k_{prod} E_{prod}.$$

- The degradation of $X$ is represented by:

$$\frac{dX(t)}{dt} = -X(t) k_{deg} E_{deg}.$$

- The reversible dimerization of $X$ and $Y$ (i.e., the formation of the dimer $XY$), by:

$$\frac{dXY(t)}{dt} = X(t)Y(t) k_{dim} E_{dim} - XY(t) k_{dis} E_{dis}.$$

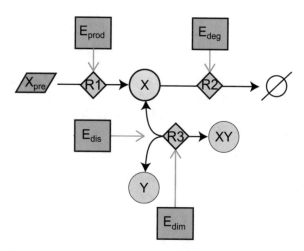

**Fig. 2.19** Example of a kinetic reaction network. The reaction network is given by the interactions between the time-varying biochemical species $X$, $Y$, and the heterodimer $XY$ (shown in gray circles). The concentrations of these species are determined by the reactions $R1$: de novo production of $X$ from a non-rate-limiting precursor $X_{pre}$; $R2$: degradation of $X$, and $R3^{+/-}$: and reversible formation of the heterodimer $XY$. These reactions are catalyzed by the enzymes $E_{prod}$, $E_{deg}$, and $E_{dis}$, and $E_{dim}$ (blue squares), whose concentrations are assumed to remain constant. Figure taken from [120] (URL: http://hdl.handle.net/10044/1/47969, published under a Creative Commons Attribution Non-Commercial No Derivatives Licence https://creativecommons.org/licenses/by-nc-nd/3.0/)

Let us remind again that effectors are distinguished from precursors in that their concentrations are not affected by the reaction they catalyze (i.e., $dE_{reaction}/dt$ is independent of these reactions).

Negative regulation is often represented in a phenomenological way, by multiplying the rate on which the repressor is acting by a function that decreases monotonically with the concentration of the repressor. It would also be possible to derive these sorts of functions from basic biochemistry, by explicitly representing, again using the Law of Mass Action, the mechanism by which the repressor exerts its action (depending on how the repressor acts: for instance, trapping the effector molecule). However, for convenience (less parameters and equations) this level of mechanistic detail is often omitted.

One of the main advantages of the mathematical description of a reaction network is that *all* the reactions that affect $X$ can be represented and studied simultaneously (this is why it is called a *systems biology approach*). Thus, if for example, $X$ is being produced, degraded, and also forms a heterodimer with another molecule $Y$ (reaction network depicted in Fig. 2.19), then its full dynamics are described by simply adding up the individual reactions explained above. In the particular example shown in Fig. 2.19, this procedure would retrieve the expression that describes $X(t)$ as:

$$\frac{dX(t)}{dt} = \overbrace{X_{\text{pre}}k_{\text{prod}}E_{\text{prod}}}^{\text{R1: de novo production}} - \overbrace{X(t)k_{\text{deg}}E_{\text{deg}}}^{\text{R2: degradation}} - \overbrace{X(t)Y(t)k_{\text{dim}}E_{\text{dim}}}^{\text{R3+: dimer formation}} + \overbrace{XY(t)k_{\text{dis}}E_{\text{dis}}}^{\text{R3-: dimer dissociation}}.$$
(2.15)

The dynamics of $X$ given by Eq. (2.15) depend on *constant parameters*, such as $k_{\text{deg}}$, $k_{\text{dim}}$, and $k_{\text{dis}}$, but also on other, time-varying variables, such as $Y(t)$ and $XY(t)$. Hence, to mathematically analyze the behavior of $X(t)$, we need also to consider the equations for $Y(t)$:

$$\frac{dY(t)}{dt} = - \overbrace{X(t)Y(t)k_{\text{dim}}E_{\text{dim}}}^{\text{R3+: dimer formation}} + \overbrace{XY(t)k_{\text{dis}}E_{\text{dis}}}^{\text{R3-: dimer dissociation}},$$
(2.16)

and for $XY(t)$:

$$\frac{dXY(t)}{dt} = \overbrace{X(t)Y(t)k_{\text{dim}}E_{\text{dim}}}^{\text{R3+: dimer formation}} - \overbrace{XY(t)k_{\text{dis}}E_{\text{dis}}}^{\text{R3-: dimer dissociation}} = -\frac{dY(t)}{dt}.$$
(2.17)

Collectively, the set of coupled equations (2.15)–(2.17) that describe all the inter-dependent variables of the system form the system of differential equations describing the reaction network.

More generally, our example is an Initial Value Problem of the form:

$$\frac{d\bar{x}(t)}{dt} = \bar{f}(\bar{x}(t), \bar{P})$$
(2.18a)

$$\bar{x}(0) = \bar{x}_0$$
(2.18b)

where:

- $\bar{x}(t)$ represents the $n$-th dimensional vector of $n$ model variables ($\bar{x}(t) = (X(t), Y(t), XY(t))$ in our example Eq. (2.15));
- $\bar{f}$ is the $n$-th dimensional function describing the dynamics of $\bar{x}(t)$, e.g., right-hand side of (2.15)–(2.17);
- and $\bar{x}_0$ is the vector of initial conditions.

### → Step 3: Model Simplification: Identification of the Conservation Equations

Let's try to make some simplifications by identifying sets of variables (also called species in this context) that together do not change over time. In other words, we are looking for sub-sets of $\bar{x}$ that satisfy:

$$\sum_{j=1}^{k} \frac{dx_j(t)}{dt} = 0$$
(2.19a)

$$\rightarrow \sum_{j=1}^{k} x_j(t) = x_k^{\text{total}} = \text{constant.} \tag{2.19b}$$

Equations (2.19) are known as *conservation equations* because they describe conserved amounts of species in a system. Dynamical systems in which sets of species are related by such equations can be simplified, since Eq. (2.19) implies that (at least) one of the variables $x_\xi$ of the subset $x_k$ can be described algebraically, as a function of $x_k^{\text{total}}$ and $x_j$, $j = 1, \ldots, k-1$, i.e.:

$$x_\xi(t) = x_k^{\text{total}} - \sum_{j=1}^{k-1} x_j(t) \tag{2.20}$$

and thus, there is no need to solve (integrate) $dx_\xi(t)/dt$. The $n$-th-dimensional system (2.18) can then be simplified to an $(n-1)$-th dimensional system of ordinary differential equations with one algebraic equation (2.20).

*Remark 2.34 (Simplifications Are Not Always Possible)* Not all systems can be simplified by conservation equations; only those in which at least some of the variables are neither produced de novo nor degraded.

In our example, we can see already from the reaction network in Fig. 2.19 that although variable $X$ is being produced *and* degraded, neither the monomer $Y$ nor the heterodimer $XY$ are being produced de novo nor degraded—they are good candidates for our conservation equations. Looking at the corresponding ODEs, we can indeed show that $\frac{dXY(t)}{dt} + \frac{dY(t)}{dt} = 0$. In other words, the total amount of $Y$ is conserved (i.e., the sum of free $Y$ and heterodimer-$Y$ ($XY$) is constant), and thus, defining $Y_T$ as the total amount of $Y$ ($Y_T = Y + XY$):

$$\frac{dY_T(t)}{dt} = \frac{dXY(t)}{dt} + \frac{dY(t)}{dt} = 0 \rightarrow Y_T = Y(t) + XY(t) = \text{constant.} \tag{2.21}$$

These conservation equations imply that $Y(t) = Y_T - XY(t)$ (or, equivalently, $XY(t) = Y_T - Y(t)$) and thus, there is no need to solve (integrate) $\frac{dY(t)}{dt}$ (or $\frac{dXY(t)}{dt}$) to obtain $Y(t)$ (or $XY(t)$). The three-dimensional system of ordinary differential equations given by the coupling of Eqs. (2.15)–(2.17), can then be simplified to the two-dimensional system given by:

$$\frac{dX(t)}{dt} = X_{\text{pre}} k_{\text{prod}} E_{\text{prod}} - X(t) k_{\text{deg}} E_{\text{deg}} - X(t) Y(t) k_{\text{dim}} E_{\text{dim}} + (Y_T - Y(t)) k_{\text{dis}} E_{\text{dis}},$$
$$\tag{2.22}$$

$$\frac{dY(t)}{dt} = -X(t) Y(t) k_{\text{dim}} E_{\text{dim}} + (Y_T - Y(t)) k_{\text{dis}} E_{\text{dis}}. \tag{2.23}$$

### → Step 4: Identification of the Initial Conditions, Parameters, Experimental Data and Qualitative Behaviors to be Reproduced by the Model

Before proceeding to the analysis of the system, it is necessary to gather all the relevant experimental information to which the model simulations will be compared. Although this is an important step in all types of mathematical modeling, it is particularly important for mechanistic ordinary differential equations, since their behavior can be strongly affected by the chosen model parameters. Indeed, while the behavior of the Boolean network models we revised earlier in this book (namely, number, composition, and topology of the attractors) depend solely on the network structure (i.e., there is one and only one behavior per Boolean model), depending on the parameter choice a single ODE-based model can display significantly different behaviors. For example, the model of the NF$\kappa$B response pathway proposed in [221] can show either an oscillatory or a bistable behavior. Also, the model of *Atopic dermatitis* (see next chapter), can display:

- monostability,
- bistability,
- or oscillations.

Thus, prior knowledge on the initial conditions, critical parameters, experimental data, and expected qualitative behaviors to be reproduced by the model is very useful to constrain the analysis of the system of ordinary differential equations. For example:

- What is the typical range of concentrations/numbers in which we expect to find the model variables?
- How is the input–output relation? Equivalently, are dose–response curves available?
- Is there any critical parameter known to drastically change this input–output response (mutations, further environmental factors, etc.)?
- How is the dynamic response of the measurable output to changes in the input?
- What is the time-resolution of typical experiments that empirically describe the system?

### → Step 5: Finding the Equilibrium Behavior of the System: Steady-State Analysis

We will start the mathematical analysis of our system of ODEs by identifying the steady-state behavior of the system (i.e., the circumstances of equilibrium). Per definition, when a system is in steady state, its rate-of-change is equal to zero. These *steady-state values* $\bar{x}_{ss}$ satisfy that:

$$\bar{x}_{ss}(t) = \bar{x}_{ss}(t + \Delta t) \ \ \forall \Delta t > 0$$

(equivalent to the *attractors* of the Boolean networks). These conditions are fulfilled when the rate of change of $\bar{x}$, given by (Eq. (2.18)):

$$\frac{d\bar{x}(t)}{dt} = \bar{f}\left(\bar{x}(t), \bar{P}\right),$$

is equal to zero, i.e., if $\bar{x}_{ss}$ is a steady-state value then:

$$\frac{d\bar{x}_{ss}}{dt} = 0 \rightarrow \bar{f}(\bar{x}_{ss}, \bar{P}) = 0.$$

To obtain the steady-state value(s) $\bar{x}_{ss}$, it thus is necessary to solve the algebraic equation:

$$\bar{f}(\bar{x}_{ss}, \bar{P}) = 0 \tag{2.24}$$

For some systems of equations, it is possible to analytically derive an expression for $\bar{x}_{ss}$ as a function of parameters. i.e., by solving (2.24), one can obtain the function $G(\bar{P})$ such that:

$$\bar{x}_{ss} = G(\bar{P}). \tag{2.25}$$

For example, the steady-state behavior $[X_{ss}, Y_{ss}]$ of our reaction network depicted in Fig. 2.20 and described by the system of equations (2.23) is obtained by simultaneously solving: $\frac{dX(t)}{dt} = 0$ and $\frac{dY(t)}{dt} = 0$, leading to:

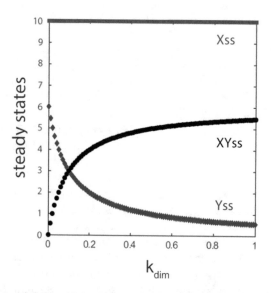

**Fig. 2.20** Steady-state behavior of the example kinetic reaction network. This figure shows the steady-state behavior of the example kinetic reaction network as a function of the dimerization constant $k_{dim}$. Increasing the dimerization constant $k_{dim}$ leads to a monotonous decrease in $Y_{ss}$ (Eq. (2.26b)), mirrored by an increase in $XY_{ss}$. The values of $X_{ss}$ are unaffected by changes in this parameter (Parameter values: $X_{pre} = 10, k_{prod} = 1, E_{prod} = 0.5, k_{deg} = 1, E_{deg} = 0.5, E_{dim} = 10, k_{dis} = 1, E_{dis} = 1, Y_{tot} = 6$ and $k_{dim} = [0 : 0.01 : 1]$)

$$X_{ss} = \frac{E_{prod}\,X_{pre}\,k_{prod}}{E_{deg}\,k_{deg}}, \tag{2.26a}$$

$$Y_{ss} = \frac{E_{deg}\,E_{dis}\,Y_{tot}\,k_{deg}\,k_{dis}}{E_{deg}\,E_{dis}\,k_{deg}\,k_{dis} + E_{dim}\,E_{prod}\,X_{pre}\,k_{dim}\,k_{prod}}. \tag{2.26b}$$

While the expressions for $X_{ss}$ and $Y_{ss}$ can be easily obtained by hand, it is often useful to use software to obtain such steady-state values. Different types of software can be used for this, as long as it has the possibility to preform symbolic computations (e.g., Mathematica, Maple, Macsyma, etc.).

In Box 2.1, we exemplify how Matlab can be used to obtain expressions (2.26).

---

**Box 2.1.** Using software to compute analytical expressions of steady states

```
1  % (1) declare parameters as symbolic variables
2  syms Xpre kprod Eprod kdeg Edeg kdim Edim kdis
       Edis Ytot
3  % ... and also the variables
4  syms X Y
5
6  % (2) write the equations:
7  XY=Ytot-Y;
8  dXdt=Xpre*kprod*Eprod-X*kdeg*Edeg-X*Y*kdim*Edim+XY
       *kdis*Edis;
9  dYdt=-X*Y*kdim*Edim+XY*kdis*Edis;
10
11 % (3) obtain the steady states
12 [Xss, Yss]=solve([dXdt==0, dYdt==0], [X, Y]);
```

---

These expressions (Eq. (2.23)) can be used to assess the parameter dependency on the long-term behavior of the system. For example, Fig. 2.20 shows the steady-state behavior of the example kinetic reaction network as a function of the dimerization constant $k_{dim}$ from 0 to 1 while keeping the other parameter values constant ($X_{pre} = 10$, $k_{prod} = 1$, $E_{prod} = 0.5$, $k_{deg} = 1$, $E_{deg} = 0.5$, $E_{dim} = 10$, $k_{dis} = 1$, $E_{dis} = 1$, $Y_{tot} = 6$). Increasing the dimerization constant $k_{dim}$ leads to a monotonous decrease in $Y_{ss}$ (Eq. (2.26b)), mirrored by an increase in $XY_{ss}$. The values of $X_{ss}$ are unaffected by changes in this parameter.

In general, neither the existence nor the uniqueness of a steady state can be guaranteed. For example, consider the simplest, zero-order ODE describing the constant production of $B(t)$ ($\emptyset \rightarrow B$ in our graphical notation given in Fig. 2.18) $\frac{B(t)}{dt} = a$, for $a > 0$. This equation has no steady-state solution (i.e., there is not $B_{ss}$ such that $\frac{dB_{ss}(t)}{dt} = 0$), which makes sense, since we are considering a constant production of $B$. In the other extreme, there are also many biologically relevant systems that (depending on parameter choices) can display *multiple* steady states. As in Boolean network models displaying multiple attractors, such multistable systems can be used to represent the phenotypic plasticity displayed by biological systems, and are hence a particularly important class of ordinary differential equations models.

The following Box 2.2 is devoted to its description.

---

**Box 2.2.** Bistability—fragmentation of the phenotype space

*Definition of Bistability*
A switch-like dose–response behavior refers to the relation between a input (commonly, a ligand) and the steady-state concentrations of an output, where small changes in the input can drive large changes in the output (see for instance [171]). A particular class of such a switch-like behavior is *bistability*, in which this abrupt change in output concentration is also *history-dependent*. In such a bistable dose–response behavior, the critical concentration of the input at which the abrupt switching *onset* from low to high values occurs is different from the critical input concentration that triggers the *ceasing* of the switch, back from high to low values. The region between the two threshold values for cease and onset of the output response is termed *bistable region* because the output can have two possible values, high or low, depending on the *previous* values of the output; if previous values are low, then the system remains at the low branch, and vice versa. This property confers the system with *memory*, also termed as *hysteresis*, since the current state depends on past values (see Fig. 2.21).

*Functional Implications of Bistable Biological Systems*
For cellular systems, the existence of bistability (or multistability in general, as discussed in previous sections of this chapter) has enormous functional implications. If a state of a cell is interpreted as a phenotype, then the multistability of a cellular system corresponds to the spectrum of different phenotypes that can be attained by a particular cell with a particular *reaction network configuration*. Each of these states, or phenotypes, has an associated *basin of attraction*, the size of which is related to the stability and the robustness to stochastic, intrinsic (e.g., genetic) and environmental perturbations (see Fig. 2.22). For example, many small signaling networks controlling abrupt, all-or-nothing phenotypes have been modeled with ordinary differential equations, which result in bistability. Examples include apoptosis [129, 194]; cell cycle progression [513]; commitment to meiosis [125]; oocite maturation [147, 148]; quorum sensing in bacteria [482], and immune responses elicited by dendritic cells [413], keratinocytes in psioratic [461] and atopic dermatitis [439] lesions, T cells [201, 279], lymphocytes [512], chondrocytes [343], macrophages [432] and endothelial cells [332], among others. In general, computational [412, 451, 489] and theoretical

---

(continued)

**Box 2.2.** (continued)

[22, 101] analysis of these and other biochemical networks have shown that bistability can result from biochemical networks displaying positive feedback with cooperativity.

*Assessing the Effects of Genetic and Environmental Risk Factors*
Using the genetic deficiency as a bifurcation parameter, it is possible to systematically assess how the properties of the basin of attraction of a particular cell state are affected by the strength of the genetic deficiency [22] (analogously to the analysis of the changes in the epigenetic landscape elicited by genetic perturbations, as schematically shown in Fig. 2.22b). External (environmental) perturbations are required to force the system from one state to the other, crossing the sepparatrix that divides the basins of attraction (Fig. 2.22b). With a mathematical model one can assess how the *minimal magnitude or duration* of an external challenge required to drive a phenotype transition is affected by genetic perturbations (Fig. 2.22c). Such a *susceptibility* analysis can be used to characterize how disease progression emerges from the complex interplays between genetic and environmental risk factors (Fig. 1.1), as will be exemplified in the last chapter.

*Complexity of Discrete vs Continuous Multistable Models*
It is important to point out that due to the theoretical and computational constraints of the current methods to analyze systems of nonlinear differential equations (which give rise to multistability), in general a nonlinear ordinary differential equations approach to analyze multistability should be favored over a Boolean approach only if the system under study and the number of attractors or phenotypes to be described are small enough, and if there is an explicit need to analyze the network assuming continuous-time and continuous-state variables. In principle, to determine if an ordinary differential equations system shows bistability, it is enough to solve Eq. (2.24) and find that there are three (two locally stable, one locally unstable; stability can be determined by the analysis of the corresponding Jacobian matrix) steady-state solutions. However, in practice, most of the ordinary differential equations systems that can show bistability are highly nonlinear (indeed, nonlinear positive feedback interactions are required for such a qualitative behavior, see for instance [22, 101]), and thus, the solution to Eq. (2.24) can seldom be determined analytically. Numerically, of course one could either integrate the ordinary differential equations from varying initial conditions (discussed in the next section), or compute the different roots of Eq. (2.24) (e.g., using the Newton-Raphson algorithm). The difficulty lies on the fact that one cannot know a priori if a given ordinary differential equation system with a specific parameter set shows bistability, even if it displays the structural features for this behavior. Thus, for each parameter set, multiple initial conditions (numerical integration) or initial guesses

---

**Box 2.2.** (continued)

(numerical determinations of the roots of the algebraic equation) have to be tested. To address these issues, several software toolboxes have been proposed for the construction of such bifurcation diagrams. Examples are Matcont or the Dynamical Systems Toolbox, both for Matlab; Oscill8; XPPAUT; GRIND and COPASI. All of these programs use *numerical continuation algorithms* to computationally approximate the long-term behavior of the nonlinear ODEs as a function of a bifurcation parameter [139, 256]. Such results can be graphically represented by bifurcation diagrams, which in a bistable system describe the abrupt switching between two stable steady states when the bifurcation parameter reaches its *cease* or *onset* thresholds (Fig. 2.21).

---

### → Step 6: Dynamical Simulation of the System of Ordinary Differential Equations

When we formulate an ordinary differential equations based mechanistic model, such as in the coupling between Eqs. (2.15), (2.17), and (2.16), what we describe are the *rates of change* of the dynamic variables, i.e., $d\bar{x}(t)/dt$ with $\bar{x}(t) = [X(t), Y(t), XY(t)]$. But wait! When looking for the system's dynamics, what we actually want to know is not exactly $d\bar{x}(t)/dt$, but rather $\bar{x}(t)$. So, what we have to do is to *deduce* $\bar{x}(t)$ from the information that we have, i.e., from $d\bar{x}(t)/dt$. This is, we have to integrate, or solve, $d\bar{x}(t)/dt$. How? There is bad news and good news.

The bad news is that the vast majority of the biologically relevant mechanistic ordinary differential equations based models are highly nonlinear systems of ordinary differential equations describing the rate of change of more than one state variable, and in general such models do not have in general an analytical solution [429]. This means that in most cases we cannot aim to obtain a *function* $\bar{G}(t, \bar{P})$ such that $\bar{x}(t) = \bar{G}(t, \bar{P})$.

The good news is that there are many numerical methods to estimate $\bar{x}(t)$ from $d\bar{x}(t)/dt$ (and the initial conditions $x(0)$), for example, the Euler or the Runge–Kutta methods (see for instance [421]). It is beyond the scope of this book to discuss in detail how these methods work (interested readers are referred to [429] or [421]), but the basic principles are illustrated with the simplest method for numerical integration of systems of ordinary differential equations, namely the Euler method, as follows:

### Numerical Integration via the Euler Method

Consider the Initial Value Problem given by Eq. (2.18). To numerically integrate this model, one takes a current, known value:

Input (stimulus)

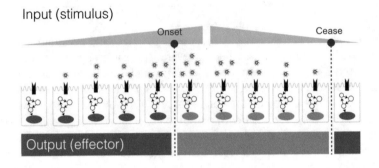

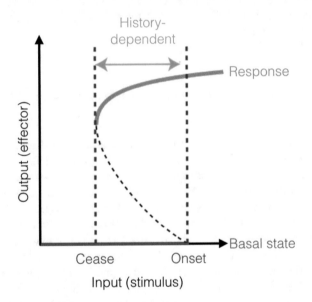

**Fig. 2.21** Bifurcation diagram of a continuous bistable switch. Schematic representation of a bistable dose–response (bifurcation) behavior describing the relation between the concentration of the input (*stimulus*) and the steady-state concentration of the output (*effector*). Effector concentrations remain at low values until a critical threshold for onset in the stimulus concentration is reached, triggering the abrupt activation of the effector. High effector values can be decreased only if the ceasing threshold is reached. The history-dependent region comprised between ceasing and onset thresholds is termed bistable region. Example of such bistable dose–response behavior is the abrupt and history-dependent onset and cease of innate immune responses that are triggered in response to pathogens that come in contact with epithelial cells

$$\bar{x}(t_0) = \bar{x}_0,$$

and using the knowledge about the expected dynamics, comprised in the derivative $\bar{f}(\bar{x}(t), \bar{P})$, one can estimate the value in the next time step:

$$t_1 = t_0 + \Delta t,$$

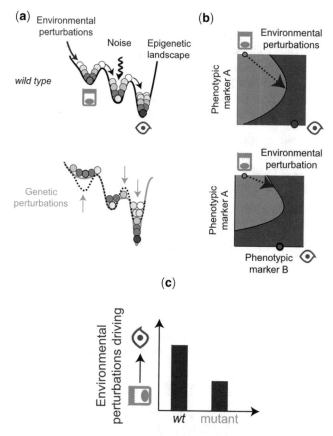

**Fig. 2.22** Genetic risk factors affect the susceptibility of changing the phenotype in response to environmental perturbations by altering the size and structure of the basins of attraction. (**a**) The topology of all the basins of attraction of a given gene regulatory network can be represented by the epigenetic landscape (discussed previously). Environmental perturbations and noise can trigger the movement between attractors (i.e., a phase shift). Genetic perturbations can directly affect the structure of the epigenetic landscape. (**b**) shows an alternate representation of the attractor landscape associated to a regulatory network: the state-space description, which maps the initial conditions to its attractors, defining the basins of attraction. As in (**a**), movement between attractors can be driven by environmental perturbations that are strong enough to force the state from one basin into the other. As shown is (**c**), genetic perturbations can ease or hinder this transition by affecting size and structure of the basins of attraction

with $\Delta_t \to 0$ a sufficiently small time step, as:

$$\bar{x}(t_0 + \Delta t) = \bar{x}_0 + \Delta t \, \bar{f}(\bar{x}(t_0), \bar{P}).$$

Repeating this procedure iteratively from $t_0$ to $t_n$, one can obtain an numerical estimation for the dynamics for $\bar{x}(t)$. The smaller the time step, the more accurate the calculation, but, if done manually, also the more cumbersome the procedure!

Fortunately, there are many software choices with built-in ordinary differential equations solvers. For example, the statistical programming language R, already mentioned in previous sections of this book, has a library (deSolve) with many functions to perform this task see [421]. Also Matlab has many built-in functions for this (for example, ode45), and extensive documentation.

Note that these solvers will implement more sophisticated integration algorithms than this simple Euler method, but the procedure remains the same:

1. Declare the ordinary differential equations function (Eq. (2.18a)).
2. Define the parameter value ($\bar{P}$).
3. Define the initial condition ($\bar{x}(t_0) = \bar{x}_0$).
4. Define the integration interval.

   Note: most of the computational ODE solvers "choose" $\Delta_t$, hence it is only necessary to set the initial and the final times.

5. Call the ordinary differential equations solver, and obtain $\bar{x}(t)$.

To illustrate this procedure, let's see how these steps can be implemented in Matlab. First, we write the system of ordinary differential equations (Eq. (2.23)) in a separate m-file. The name of the file should be the name of the function. In our case, we call it dimerFormation.m:

Note: it is also possible to declare the function within the same script where it will be solved (as an anonymous function), but it is generally a good practice to define functions in separate files.

```
1  function dydt=dimerFormation(~,y ,Xpre, kprodEprod
       , kdegEdeg, kdimEdim, kdisEdis , Y_tot)
2
3  dydt = zeros(2,1);
4
5  X_t=y(1);
6  Y_t=y(2);
7  XY_t=(Y_tot-Y_t);
8
9  dydt(1)=Xpre*kprodEprod-X_t*kdegEdeg-X_t*Y_t*
       kdimEdim+XY_t*kdisEdis;
10 dydt(2)=-X_t*Y_t*kdimEdim+XY_t*kdisEdis;
```

Now we are ready for the numerical integration:

```
1  % (1) Define the constant parametre values
2  Xpre=10; kprodEprod=.5; kdegEdeg=.5; kdimEdim=10;
       kdisEdis=1; Ytot=6;
```

(continued)

> **Box 2.2.** (continued)
>
> ```
> 3  % (2) Define the initial condition
> 4  y0=[0 Ytot];
> 5  % (3) Define the integration interval
> 6  tspan = [0 5];
> 7  % (4) Call the ODE solver
> 8  [t,y] = ode45(@(t,y)dimerFormation(t,y ,Xpre,
>        kprodEprod, kdegEdeg, kdimEdim, kdisEdis, Ytot
>        ),tspan,y0);
> ```

We can then visualize the results, obtaining in our case the dynamic trajectories for $\bar{x}(t) = [X(t), Y(t), XY(t)]$ shown in Fig. 2.23.

### → Step 7: Parameter Optimization (Seeking the Best Agreement Between the Mathematical Model and the Experimental Data, Using Minimization Algorithms)

Ordinary differential equations models are quantitative. This means that there is an explicit dependency on the transitory and steady-state behaviors of the state variables on the choice of parameter values. So, how to choose the best parameter set $\bar{P}_{opt}$?

Ideally, one should use multi-dimensional *quantitative* data (to obtain experimental values for all our dynamic variables, i.e. $\bar{x}_{exp}(t_{exp})$) to find this optimal parameter

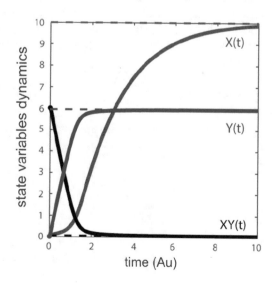

**Fig. 2.23** Dynamic behavior of the example kinetic reaction network with the settings specified in the code 2.8. The horizontal dotted lines represent the steady-state values (Eq. (2.26))

set. To do so, the idea is to find the optimal parameter set $\bar{P}_{\text{opt}}$ such that the model $\bar{x}(t) = \bar{f}(t, \bar{P}_{\text{opt}})$ matches as closely as possible the experimental data $\bar{x}_{\text{exp}}$. In other words, we are looking for the parameter set $\bar{P}_{\text{opt}}$ that *minimizes* the difference between the experimental data and the model, i.e., the *cost* given by:

$$\text{cost}(\bar{P}) = \sum_{i=1}^{k} (\bar{x}_{\text{exp}}(t_i) - \bar{x}(t_i, \bar{P}))^2 \tag{2.27}$$

To find the solution to this minimization problem:

$$\min(\text{cost}(\bar{P})) = \text{cost}(\bar{P}_{\text{opt}}) \tag{2.28}$$

one can use nonlinear optimization algorithms such as the Nelder–Mead simplex algorithm. These can be implemented by most of the software with nonlinear ordinary differential equations analysis capabilities. Below we will give an example of how this algorithm can be implemented in Matlab, by using the built-in function fminsearch (Box 2.3). In our example, we will seek the optimal agreement between simulations of our reaction network representing the regulatory interactions controlling the formation of the heterodimer $XY$ (Eq. (2.23)) and the experimental data from [417] (Fig. 5B). This data describes the binding of the activated transcription factor Smad1 (corresponding to our variable $X$; since Smad1 can be activated/produced, degraded, and forms a heterodimer) to the promoter of the PPAR$\gamma$ gene (corresponding to our variable $Y$, since DNA sequences such as promoters are neither produced nor degraded in this time-scale (hours); their monomeric, free concentrations are only affected by dimerization with regulatory proteins such as transcription factors), in response to stimulation with the extracellular ligand BMP. For our optimization, we will assume that the parameter values $X_{\text{pre}} = 10$, $k_{\text{dis}} = k_{\text{prod}} = k_{\text{deg}} = 0.5$, $E_{\text{dim}} = E_{\text{prod}} = E_{\text{dis}} = 1$ are fixed (this could be justified if these parameter values were calculated from other empirical data). We will find the optimal $\bar{P}_{\text{opt}}[k_{\text{dim}}, Y_{\text{tot}}]$ parameter pair that best reproduces the dataset.

**Box 2.3.** Numerical optimization with Matlab

```
1  function xOPT=example_optimization_dimerFormation
2  %% Experimental data
3  t_exp= [0 1 3 ]; % hours post-stimulation
4  XY_exp=[.5 2.5 3 ]./.5; %Smad1- promoter complex
5  %% Optimization
6  % Give an initial guess for the parameter
7  xinit= [2 7];
8  % Run the optimization!
```

(continued)

---

**Box 2.3.**   (continued)

```
 9  [x,J,flag]=fminsearch(@(x)CostFunction(x,t_exp,
        XY_exp),xinit);
10  xOPT=x;
11  end
12  function Cost=CostFunction(x,t_exp, XY_exp)
13  kdimEdim=x(1);
14  Ytot=x(2);
15  %% Solve the ODE - with this parameter
16  %constant parameters (those not minimized)
17  Xpre=10; kprodEprod=.5; kdegEdeg=.5; kdisEdis=1;
18  %initial conditions
19  X_0=0; Y_0=Ytot-XY_exp(1);
20  y0=[X_0 Y_0]; %initial condition [X, Y]
21  %integration interval
22  tspan = [0 t_exp(end)]; %integration interval
23  % Call the ODE solver
24  [t,y] = ode45(@(t,y)dimerFormation(t,y ,Xpre,
        kprodEprod, kdegEdeg, kdimEdim, kdisEdis, Ytot
        ),tspan,y0);
25  % Focus on the variable to be compared with data
26  XY_t=Ytot-y(:,2);
27  % interpolate those values corresponding to the
        measurements
28  XY_predicted = interp1(t,XY_t,t_exp);
29  %% Calculate the cost of the predition vs. the
        experimental data
30  Cost=(sum((XY_predicted-XY_exp).^2));
31  end
```

---

We can then visualize the results of our optimization by calculating $\bar{P}_{opt} = [k_{dim}, Y_{tot}]$ with our optimization function, and then plotting the model $\bar{x}(t, \bar{P}_{opt})$ together with the experimental data used for the optimization, as shown in Box 2.4.

---

**Box 2.4.**  Visualization of the optimal solution

```
1  close all; clear all; clc
2  %% Run the ODE with the optimal parameters
3  % constant parameter values
4  Xpre=10; kprodEprod=.5; kdegEdeg=.5; kdisEdis=1;
5  % minimized parameter values
```

---

**Box 2.4.**   (continued)

```
6   xOPT=example_optimization_dimerFormation;
7   kdimEdim=xOPT(1); Y_tot=xOPT(2);
8   % experimental data
9   t_exp=  [0 1 3 ]; XY_exp=[.5 2.5 3 ]./.5;
10  % Initial conditions
11  X_0=0; Y_0=Y_tot-XY_exp(1);
12  y0=[X_0 Y_0];
13  % integration interval
14  tspan = [0 t_exp(end)+.5];
15  % Call the ODE solver
16  [t,y] = ode45(@(t,y)dimerFormation(t,y ,Xpre,
        kprodEprod, kdegEdeg, kdimEdim, kdisEdis,
        Y_tot),tspan,y0);
17
18  %% calculate the final cost
19  XY_pred = interp1(t,(Y_tot-y(:,2)),t_exp);
20  Cost=(sum((XY_pred-XY_exp).^2));
21
22  %%Plot the results
23  figure;
24  scatter(t_exp, XY_exp, 'k'); hold on
25  plot(t, Y_tot-y(:,2), 'k'); axis square
26  ylabel('XY dynamics [fold increase]')
27  xlabel('time [hours]')
28  title([ 'minimal cost:' num2str(Cost) ' kdim='
        num2str(xOPT(1)) ' Ytot=' num2str(xOPT(2))]);
```

These results are shown in Fig. 2.24.

Let us conclude this small discussion on parameter optimization with a word of caution.

**Some Comments on Parameter Optimization**

- Over-fitting of parameters might occur if the ratio of parameters to be optimized relative to high-quality experimental information is unfavorable. Thus, the more coherent (i.e., from the same experiment, ideally) empirical data we have for the parameter optimization, the better. It is important to aim for a (parametrized) model that *robustly* reproduces the expected behaviors [277, 426].
- The optimal solution $\bar{P}_{opt}$ might not be unique. In fact, a practical problem when searching for $\bar{P}_{opt}$ is that the minimization algorithm can be trapped in a local

**Fig. 2.24** Example parameter optimization. Optimal agreement between mathematical model (Eq. (2.23)) and experimental data given in [417]. The obtained optimal parameters are $\vec{P}_{opt} = [k_{dim} = 8.5, Y_{tot} = 6.1]$, which result in the minimal cost (Eq. (2.27)) $2.5 \times 10^{-11}$—That's small indeed!

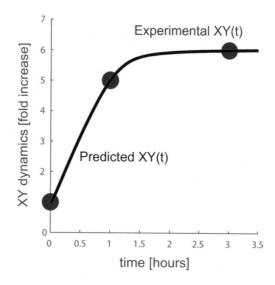

minimum, i.e., where the resulting cost is low but not the (globally) lowest. To avoid these complications, whenever possible, a global optimization algorithm (e.g., simulated annealing [247]) should be preferred over local minimization algorithms such as `fminsearch`.

- Other, simpler (with less variables and parameters) models might be able to better explain or reproduce the experimental data. If in doubt over the regulatory interactions underlying the behavior of the model variables, it is advisable to representing a by proposing a set of plausible models (with different kinetic reactions representing a different network topology), and systematically testing how well these models can fit the experimental data. Then, one can use the Akaike Information Criterion or other statistical techniques to select the simplest model able to best reproduce the experimental data (see the details in [453]).

In conclusion, parameter optimization is a powerful tool that can help to find the parameter set with which the proposed model can best describe a given set of experimental data. Often, this technique is used simply as a methodological step, to parameterize the ordinary differential equations model for further mathematical analysis. However, as more high-throughput and quantitative experimental data becomes available, parameter optimization routines can be used to directly address clinically relevant research problems. For example, parameter optimization has been used to deduce from data the underlying cause of a pathogenic transformation of the liver [195, 449, 450, 464], and helped to stratify patient cohorts in different personalized treatments groups [149, 187, 395].

We are ready now to conclude our procedure.

→ **Step 8: Model Analysis (Assessing the Robustness/Plasticity of the Model Behavior in Response to Perturbations)**

Once we have an experimentally calibrated, mechanistic model, we can start with the analysis. In most cases, clinically relevant research questions can be rephrased as:

How do structural (i.e., in the model equations) or quantitative (i.e., in the model parameters) *perturbations* affect the behavior of the model?

which is equivalent to asking:

How do genetic and/or environmental risk factors or treatment combinations affect the patient outcome (see Figs. 1.1 and 1.4)?

To answer these questions, we first have to ask:

Which *feature* of our model are we interested in analyzing, in terms of its robustness?

For example:

- Is it the steady-state value?
- Is it the existence of multiple equilibrium points?
- The existence of oscillations?
- The frequency or amplitude of the transient response?
- The time-to-relaxation?

It is important to be specific about these features, since in order to assess their robustness we must be able to bring them to formal terms. Once we have identified the feature, we can proceed to the analysis of its robustness.

*Remark 2.35* Note that both structural and quantitative features can be modulated by parametric variations (this is because the presence or absence of a specific rate, i.e. a structural change, can be represented by setting the corresponding kinetic parameter to a value $> 0$ or $= 0$).

Usually, this analysis is done by:

**Robustness analysis:**   Randomly varying all the model parameters around the nominal (optimized) value, and assessing the response of the system to these perturbations, for example, by computing the proportion of parametric model variants displaying the desired behavior. See for example the robustness analysis of a host–commensal bacteria interaction network reported in [121].

**Sensitivity analysis:**   Systematically varying all the model parameter combinations, and evaluating which parametric changes are responsible for the largest deviation in the model behaviors [74, 301].

**Bifurcation analysis:**   Follow the model's behavior in response to changes in a subset of parameters—the bifurcation parameters. Generally, this type of analysis is performed when a sharp, qualitative transition in response to a small, quantitative change in a bifurcation parameter is expected, for example, when looking for a hysteretic switch as depicted in Fig. 2.21 (see for example [123, 438]).

It is time to discuss the multiple time-scales modeling issue.

## 2.9   Modeling the Interplay Between Regulatory Networks and the Microenvironment

We explore in this section some mathematical tools intended to analyze biological dynamical systems that evolve over several time-scales. Specifically, we consider the interplay between biological processes occurring at two time-scales:

**Fast processes:**   Biochemical processes that regulate the phenotypic decision-making of cells in response to microenvironmental conditions.

**Slow processes:**   Tissue level processes regulating the dynamics of the microenvironment.

As discussed previously, addressing this issue is clinically relevant, since the characteristic gradual aggravation of chronic degenerative diseases often emerges from aberrations in the phenotype–microenvironment interactions (Figs. 1.3 and 1.4).

In previous sections, we saw that phenotypes can be mathematically represented as attractors of the underlying regulatory networks, and that transitions between these attractors can be driven not only by stochastic fluctuations, but also by changes in the microenvironmental conditions. Let us pose now the following questions:

- What if these environmental fluctuations are changing *as a consequence* of the phenotype changes driven by the individual cells in the tissue (see Fig. 1.3)?
- How to account for tissue-level risk factors, which might propagate across this multi-scale regulatory network, giving rise to the gradual phenotypic deterioration (see Fig. 1.4)?

To model this kind of systems, we will simultaneously consider:

- The changes in the activation state of biochemical reaction networks controlling phenotypic decisions.
  and:
- The tissue-level processes underlying microenvironmental fluctuations.


### *Time-Scale Separation*

While the biochemical reactions are fast, in the time-scale of minutes to hours, the dynamics of the surrounding tissue-level conditions stabilize within days to weeks. To account for these two different time-scales, we will perform a *time-scale separation*, also known as Quasi-Steady-State Assumption (QSSA), as presented in Box 2.5.

**Box 2.5.** Quasi-Steady-State Assumption (QSSA)

Assume that $S(t)$ is a slowly changing micro environmental condition, and that $X(t)$ is the cellular phenotype. Assume further that the dynamics of $S$ and of $X$ are coupled, i.e. $\dot{S} = F(S, X)$ and $\dot{X} = G(S, X)$.

Then, if we assume that the dynamics of $X(t)$ are much faster as those of $S(t)$, then $\dot{X} = 0$. This means that the relation between the microenvironmental factor $S$ and the phenotype $X$ is described algebraically, by the mapping of the bifurcation parameter $S$ to the stationary solution $X_{ss}(S)$ of equation

$$\dot{X} = G(S, X) = 0.$$

The bifurcation parameter $S$, in turn, is dynamically described by:

$$\dot{S} = F(\tau, X_{ss}(S)),$$

with $t$ and $\tau$ the time-scales of the fast and the slow system, respectively.

*Remark 2.36 (The Bifurcation Parameter Under the Influence of the Proportion of Phenotypes within the Tissue)* Since the governing function $F(\tau, X_{ss}(S))$ for the dynamics of $S$ (see Box 2.5) explicitly considers the algebraic variable $X_{ss}(S)$, the changes in the bifurcation parameter depend on the proportion of phenotypes within the tissue.

Assuming such differences in time-scales can greatly simplify the analysis of multi-dimensional systems described by the coupling between $\bar{X}$ and $\bar{S}$, with $\bar{X}$ and $\bar{S}$ being $n$ and $m$ dimensional vectors, respectively. To illustrate this, let's consider now an example.

### The Brigss–Haldane Dynamical System

The typical example of a biochemical network described by a system of ordinary differential equations and simplified by the QSSA is the Briggs–Haldane version of the Michaelis–Menten equation (see [64, 454] for the details).

The system of equations:

$$\frac{d[E]}{dt} = -k_f[E][S] + k_r[ES] + k_{cat}[ES], \tag{2.29a}$$

$$\frac{d[S]}{dt} = -k_f[E][S] + k_r[ES], \tag{2.29b}$$

$$\frac{d[ES]}{dt} = k_f[E][S] - k_r[ES] - k_{cat}[ES], \tag{2.29c}$$

**Fig. 2.25** Michaelis–Menten reaction. Reaction network is composed by the dynamic interactions between enzymes ($E$), substrates ($S$), enzyme–substrate complexes ($ES$), and products ($P$) of the enzymatic reaction represented in Eq. (2.29)

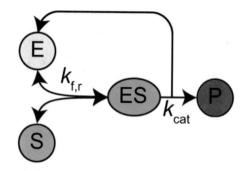

$$\frac{d[P]}{dt} = k_{\text{cat}}[ES],\tag{2.29d}$$

represents the dynamical interactions between (see Fig. 2.25):

- The catalyzing enzyme $E$.
- The substrate $S$.
- The enzyme–substrate complex $ES$.
  and:
- The product of the enzymatic reaction, $P$.

In these equations (with $k_f$, $k_r$, and $k_{\text{cat}}$, denoting reaction rates), it is considered that the total amount of enzymes is conserved (i.e., no de novo production of $E$), which can be seen directly from the conservation equation:

$$\frac{d[E]}{dt} + \frac{d[ES]}{dt} = 0,$$

which implies:

$$[E] + [ES] = [E]_0.\tag{2.30}$$

The key assumption to simplify Eqs. (2.29) is that the enzyme–substrate formation $[ES]$ is infinitely fast respect to the rest of the dynamics, i.e. $d[ES]/dt = 0$. From this QSSA (i.e., it is assumed that $[ES]$ attains its steady state $[ES]_{SS}$ in an extremely fast way), it follows that

$$k_f[E][S] = [ES]_{SS}(k_r + k_{\text{cat}}).$$

Using the conservation equation (2.30), $k_f[E][S]$ can be rewritten as:

$$k_f[E]_0[S] - k_f[ES]_{SS}[S] = [ES]_{SS}(k_r + k_{\text{cat}}),$$

from which one can isolate the variable $[ES]_{SS}$ as:

$$[ES]_{SS} = \frac{k_f[E]_0[S]}{(k_r + k_{\text{cat}}) + k_f[S]},$$

which can be used to rewrite $\frac{d[P]}{dt}$ as:

$$\frac{d[P]}{dt} = k_{cat} \frac{k_f[E]_0[S]}{(k_r + k_{cat}) + k_f[S]}.$$

Defining:

$$K_M = \frac{k_r + k_{cat}}{k_f},$$

one can recognize the simple, one-dimensional system representing the dynamics formation of a product, known as Michaelis–Menten equation:

$$\frac{d[P]}{dt} = k_{cat} \frac{[E]_0[S]}{K_M[S]}.$$

So, using QSSA we were able to reduce a four-dimensional dynamical system to a just one-dimensional ordinary differential equations!

## Bistable Dynamics

Now, back to our original problem of coupling phenotypic decisions to microenvironmental changes. Let's consider the simplest multistable system in which gradual environmental conditions drive abrupt phenotype changes, namely a bistable system (Fig. 2.21). As discussed previously, mapping the relation between the bifurcation parameter and the stable steady-state solutions can be tricky, since analytical steady solutions for high-order nonlinear systems rarely exist, and numerical methods require exhaustive explorations of the parameter space (including initial conditions) and are often stuck in local solutions. Thus, iteratively solving such multi-time-scale problems during the numerical integration of slow variables can be computationally intensive, and might often even fail to find the desired steady-state solutions. To overcome this difficulty, it is possible to phenomenologically describe the previously characterized bistable switch by a piecewise-affine (PWA) function [82]. Such PWA approximation provides a rule that maps the input (stimulus) to the output (effector) (Fig. 2.21). For example, assuming a perfect switch, the effector can be approximated by two constant values, $E_{low}$ and $E_{high}$, representing the "low" or "high" branches of the bifurcation diagram, attained at the threshold values $S^-$ and $S^+$, respectively. Now, let's consider that our bifurcation parameter, this is, the input, changes dynamically in the time-scale $\tau$. Then the relation describing how the output $E(\tau)$ is determined by the input $S(\tau)$ and by the previous output values $E(x < \tau)$ can be approximated as follows:

- If $S(\tau) < S^-$, then $E(\tau) = E_{low}$ (effector is low if the stimulus concentration is low).

- If $S(\tau) > S^+$, then $E(\tau) = E_{high}$ (effector is high if the stimulus concentration is high).
- If $S(\tau) \in [S^-, S^+]$, then:

   - if $E(x < \tau) = E_{low}$, then $E(\tau) = E_{low}$
     or
   - if $E(x < \tau) = E_{high}$, then $E(\tau) = E_{high}$,

   corresponding to the history-dependent determination of the effector value when the stimulus is in the bistable region.

More formally, these conditions can be represented by the PWA given in Eq. (2.31) (adapted from [359]):

$$
E(\tau) = \begin{cases} E_{low} & \text{if } (S(\tau) < S^-) \text{ or } \left\{ S(\tau) \in [S^-, S^+] \text{ and } E(x < \tau) = E_{low} \right\} \\ E_{high} & \text{if } (S(\tau) > S^+) \text{ or } \left\{ S(\tau) \in [S^-, S^+] \text{ and } E(x < \tau) = E_{high} \right\}. \end{cases}
$$

$$(2.31)$$

Note that Eq. (2.31) implicitly assumes two time-scales:

**Fast time-scale:**   A time-scale $t$ that governs the stabilized biochemical inter-actions that underlie the bistable dose–response behavior. These biochemical reactions can be represented by the system of ordinary differential equations $\dot{E}(t, S, \mathbf{E})$ that operates at time-scale $t$ and has a input $S$ that does not change significantly ($S(t) \approx constant$) while $E(t)$ reaches its equilibrium value (given by $E_{low}$ or $E_{high}$, respectively).

**Slow time-scale:**   A time-scale $\tau$ that determines the dynamics of the input $S(\tau)$ by the governing equation $\dot{S}(\tau) = F(\tau, S)$.

A special case of system (2.31), which is of particular interest here, occurs when the slowly changing input $S(\tau)$ is itself determined by its quickly stabilizing output $E(t)$ (and vice-versa). In such a case, also the dynamics of $S(\tau)$ (that depend on $E(\tau)$) can be descried by the PWA given in Eq. (2.32) (adapted from [359]):

$$
\dot{S}(\tau) = \begin{cases} F_{low}(S) & \text{if } E(\tau) = E_{low} \\ F_{high}(S) & \text{if } E(\tau) = E_{high}, \end{cases}
$$

$$(2.32)$$

where $F_{low}$ and $F_{high}$ are the two *governing equations* that determine the dynamics of $S$ when $E(\tau) = E_{low}$ or $E(\tau) = E_{high}$, respectively.

Accordingly, the long-term behavior of $S$ is given by the *focal points* $S_{ss}^{low}$ and $S_{ss}^{high}$, which are the steady-state values given by the solution to $F_{low} = 0$ and $F_{high} = 0$, respectively [359].

The coupling between Eqs. (2.31) and (2.32) represents a hybrid system that has been extensively discussed and analyzed in [82, 359]. The long-term behavior of the coupled variable $S(\tau)$ and $E(t)$ is determined by the relative position of the focal points $S_{ss}^{low}$ and $S_{ss}^{high}$ respect to the threshold values $S^-$ and $S^+$, as follows (Fig. 2.26):

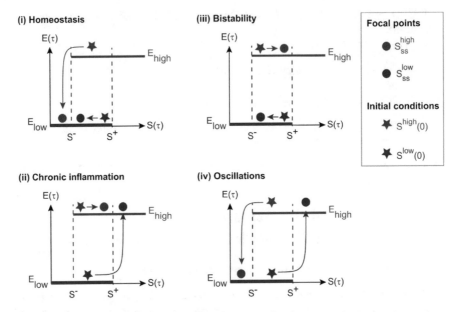

**Fig. 2.26** Schematic representation of the qualitative dynamic behaviors of the hybrid system described in the coupled equations (2.31) and (2.32). The long-term dynamical behavior of the hybrid system (2.31) and (2.32) is determined by the position of the focal points $S_{ss}^{low}$ and $S_{ss}^{high}$ respect to the threshold values $S^-$ and $S^+$. (i) If $S_{ss}^{low} \leq S^+$ and $S_{ss}^{high} < S^-$ the global equilibrium is the attractor $S_{ss}^{low}$. (ii) If $S_{ss}^{low} > S^+$ and $S_{ss}^{high} \geq S^-$, the global equilibrium is the attractor $S_{ss}^{high}$. (iii) Bistability arises from $S_{ss}^{low} \leq S^+$ but $S_{ss}^{high} \geq S^-$. (iv) Oscillations result from $S_{ss}^{low} > S^+$ and $S_{ss}^{high} < S^-$. Figure taken from [120] (URL: http://hdl.handle.net/10044/1/47969, published under a Creative Commons Attribution Non-Commercial No Derivatives License https://creativecommons.org/licenses/by-nc-nd/3.0/)

- A resting, homeostatic ("low") steady state occurs when $S_{ss}^{low} \leq S^+$ and $S_{ss}^{high} < S^-$.
- A chronically excited steady state occurs when $S_{ss}^{low} > S^+$ and $S_{ss}^{high} \geq S^-$.
- Bistability in the two time-scale dynamical system occurs when $S_{ss}^{low} \leq S^+$ but $S_{ss}^{high} \geq S^-$.
- Oscillations occur when $S_{ss}^{low} > S^+$ and $S_{ss}^{high} < S^-$.

We will see the possible clinical implications of these four qualitative behaviors in the final chapter.

## *Some Final Remarks*

This focal point analysis allows the derivation of conditions required for different qualitative behaviors of a complex dynamical system that operates in two time-scales, reducing the need for numerical methods. Note, however, that the agreement between the dynamical behavior that is analytically derived from the hybrid system representation and the numerical simulations of the model must be verified for the particular mathematical model that is analyzed using this approach to ensure that neither the discontinuities of the hybrid representation of the system nor the transient behavior that is not captured by the focal point analysis detailed above affect the dynamics of the unsimplified mathematical model.

The model concerning *atopic dermatitis* in the next chapter provides an example in which this focal point analysis is used to systematically determine the effects of risk factors affecting tissue-level processes on the development of early phases of that disease. Such framework can be applied not only to microenvironment–phenotype interactions discussed here, but in general to model (biological) systems in which there is a co-existence and inter-dependence of processes operating at different time-scales. Examples include multi-scale regulatory networks considering the interplay between:

- Metabolism and signaling [449].
- Metabolism and gene expression [359].
- Cellular-level population dynamics and biochemical processes [177, 355, 433, 434].

We can at this level discuss how to shape predictive hypothesis via the exploitation of models of regulatory networks that are relevant in the context of medical systems biology.

## 2.10 Shaping Predictive Hypothesis (Exploratory Protocol)

When tackling medical systems biology phenomena, the interplay between modeling and explicative intuition shapes predictive hypothesis. The purpose of a predictive hypothesis is to predict the nature of a relationship among the variables to be studied, which implies establishing a research agenda. This is the reason why a predictive hypothesis is also called a research hypothesis:

> *A well-thought guess, derived logically from previous findings or the predictions of a particular theory, regarding what should happen in a particular situation under certain well-defined conditions.*

Since we are concerned by chronic degenerative diseases, as resulting from the disruption of gene regulatory networks or the signal transduction mechanisms that link such networks with microenvironmental conditions, our predictive hypothesis will necessarily turn around the phenotypic consequences of such disturbances. In particular, we are motivated in using the type of systems-level modeling tools, summarized in this volume, to explore how environmental disturbances, that are related to certain lifestyles, may favor transitions from healthy to ill states. Moreover, in order to regulate disease dynamics (as discussed previously), we are interested in eventually extending the systems-level modeling approaches described here to propose preventive approaches to avoid or minimize the emergence of degenerative diseases. Once a systems-level model has been constructed for a particular disease or condition, it will be possible to proceed with the formulation of the predictive hypotheses concerning how different alterations of the regulatory networks, signal transduction pathways, or environmental factors can modulate transitions from healthy to pre-clinical, and from this to ill states. Such hypotheses should be posed so that answers can be provided to particular questions concerning such transitions. This process implies an exploratory research agenda that involves:

- Computer-based simulations.
- Data-mining.
- Epidemiological studies.
- Experiments.
- Clinical studies.

Such complete agenda for particular diseases is beyond the scope of this volume, but the models presented in the next section could be used as basic building blocks. They illustrate how the different systems-level approaches and tools presented in this chapter can be used to study:

1. Epithelial cancer.
2. Chronic inflammation.
3. Atopic dermatitis.

In each case we shall illustrate, through a bottom-up systems-level modeling approach, how the disruption of regulation gives rise to disease.

# Chapter 3
# Case Studies

## 3.1 Introduction

The aim of this chapter is to illustrate the modeling procedures discussed in the previous chapter via three examples that deal with:

| | |
|---|---|
| **EXAMPLE 1:** | Epithelial cancer |
| **EXAMPLE 2:** | Chronic inflammation |
| **EXAMPLE 3:** | Atopic dermatitis |

These three examples correspond to medical systems biology research carried out by our research team. In each case the idea is to understand how disruption of the regulatory networks leads to the onset and progression of specific chronic degenerative diseases (see Fig. 3.1). These examples illustrate how modeling can help to answer important questions concerning the systems-based mechanisms underlying chronic degenerative diseases and to characterize the interplay between the gene regulatory mechanisms and the environment. Specifically, the focus is to shed light on phenotypic plasticity.

One aim of this chapter is to illustrate how the different computational and mathematical methodologies exposed in the previous chapter of this book can be applied in practice to answer particular clinically relevant questions. For the first two case studies (epithelial cancer and chronic inflammation), we apply a qualitative discrete-time (and discrete-state) Boolean approach, whereas the third example (i.e., atopic dermatitis) follows a quantitative continuous-time and continuous-state perspective (prioritizing the unveiling of multi-scale dynamics).

The first example (i.e., epithelial cancer) is studied via discrete time and discrete state Boolean gene regulatory networks grounded on experiments. This example illustrates how the interplay between healthy, pre-clinical, and disease states, is ruled by the underlying gene regulatory machinery. A gene regulatory network module is

© Springer International Publishing AG, part of Springer Nature 2018
M. E. Álvarez-Buylla Roces et al., *Modeling Methods for Medical Systems Biology*,
Advances in Experimental Medicine and Biology 1069,
https://doi.org/10.1007/978-3-319-89354-9_3

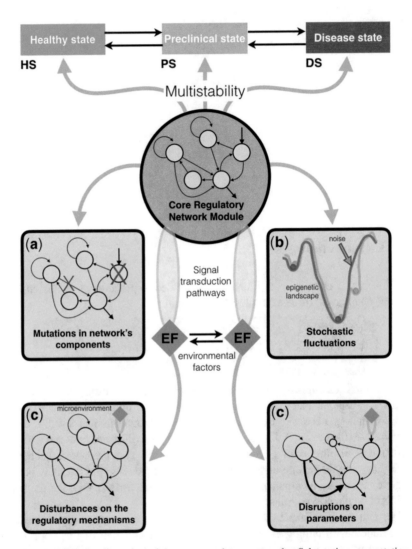

**Fig. 3.1** Modeling the disruption of the gene regulatory networks. Schematic representation of how the onset and progression of chronic degenerative diseases result from the disruption of gene regulatory networks. Transitions between healthy, pre-clinical, and disease states, corresponding to different attractors of the underlying regulatory network, are triggered by different perturbations in the form of: (**a**) Mutations in network's components; (**b**) Stochastic fluctuations; (**c**) Environmental factors; (**d**) Disruption on parameters, i.e., patient-specific network variations

proposed, and its dynamical attractors describe in systems dynamical terms the gene profiles of the:

- epithelial phenotype
- senescent phenotype
- mesenchymal phenotype

that characterize the epithelial-to-mesenchymal transition in the context of epithelial cancer. The robustness of the proposed module to genetic perturbations is analyzed through the disruption of the network's components (see Fig. 3.1a). Moreover, the epigenetic landscape are studied via stochastic simulations (see Fig. 3.1b). Our proposed bottom-up modeling approach uncovered the key role played by cell senescence in the dynamics of epithelial cancer.

In the second example, we explore how chronic inflammation can arise from the disruption of immune response by hyperinsulinemic conditions. For this, we follow a discrete-time and discrete-state Boolean gene regulatory networks modeling approach. In this case, our goal is to illustrate the dynamical consequences resulting from the interaction between the underlying core gene regulatory network module and the involved environmental factors (mediated by signaling transduction pathways), as schematically represented in Fig. 3.1c. In the context of the adaptive immune response, the example illustrates the regulatory role of the feedback-structured interplay between the intrinsic or intracellular regulatory core and the extrinsic microenvironment. Thus, the resulting Boolean network includes the gene regulatory machinery, the involved signaling pathways and their regulators, as well as cytokines that have shown to be fundamental in can help to explain CD4+ T-cell-type attainment.

A third case study illustrates how quantitative, mechanistic models based on kinetic interactions shaping the onset and progression of atopic dermatitis. For this, multi-scale, mechanistic kinetic models, constructed with ordinary differential equations, were used to systematically assess the dynamic interplay between coupled biochemical and tissue-level processes that underlie epidermal function in health and disease. We show how the model predictions are validated with clinical and experimental in vivo data. Further, we illustrate how we use such models to design therapeutic strategies to prevent or revert severe symptoms, using control theory approaches.

## 3.2 Epithelial Cancer

### *Motivation*

Cancer is a complex chronic degenerative disease that continues to challenge public health systems in poor and rich countries. Moreover, the latter have experienced increased rates of cancer incidence pointing to the important impact of environmental conditions in this complex health condition. In this section we deal with a particular class of epithelial cancers. For this we use a dynamical state-space model to exemplify a medical systems biology approach that considers cancer as a robust developmental process. Through modeling we show that cancer can emerge from altered systemic mechanisms or gene regulatory networks at play during normal development. Further, given that this and other chronic degenerative diseases have been associated to aging and to chronic inflammation, we use our model to evaluate the effects of chronic inflammation on carcinogenic transformation, and indeed show that chronic inflammation contributes to this pathogenic process.

Irrespective of the genetic background, a convergence to a finite and conserved set of phenotypes has been observed, both in physiological and pathological conditions. This convergence emerges from common and robust developmental mechanisms that restrict tissue, organ, and whole organism phenotypes as a result of the multi-level regulatory networks that encompass the dynamic interplay between gene regulatory networks, signaling molecules that respond to the microenvironment, and the environmental factors to which a person is exposed depending on the individual's lifestyle. For example, diet, as a key component of lifestyle, has been shown to have a profound influence on aging and inflammation, and it is increasingly being taken into account in scientific and clinical studies of age-related degenerative diseases (see for instance [287] and the references therein).

*Remark 3.1 (Lifestyle and Prevention)* The modulation of lifestyle and the environment may constitute an effective way to prevent and ameliorate malignancies, independently of the genetic background of an individual.

In this section we exemplify how to apply the medical systems biology framework explained in the previous chapters of this volume to understand the emergence and progression of epithelial cancer. Pathological observations of precancerous and cancerous patients indicate that cellular senescence resulting from chronic inflammation and during aging are often required to transition to a cancerous state with mesenchymal characteristics. Our proposal could aid the development of novel and more effective strategies to prevent, delay, or temporally modulate the transition to such cellular and tissue-level conditions. Indeed, carcinomas show conserved patterns of cellular behavior, and a generic time-ordered sequence of progression or transitions from certain cellular and tissue-level conditions to others are robustly displayed in most cases. This indicates equally robust underlying regulatory mechanisms that we aim to uncover (through a bottom-up state-space modeling approach). The gene regulatory network model presented here seems to constitute one of such underlying mechanisms. It incorporates components and experimentally grounded interactions that have been related to:

- progression of cell cycle
- inflammation
- epidermal and mesenchymal cell differentiation
- epigenetic regulation.

Our analysis suggests that we have uncovered a *core regulatory module* underlying epithelial-to-mesenchymal transition in which an intermediate senescent cellular state is necessary for such transition. The recovered time-ordered pattern robustly emerges from the uncovered regulatory core module.

## Cancer as a Developmental Process

Following a medical systems biology perspective, we study here cancer as a developmental process that emerges from the regulatory networks that underlie normal cell differentiation and morphogenesis during normal development. We are concerned then by

- initiation of cancer
- promotion of cancer
- progression of cancer

We focus on cancerous alterations that originate in epithelial tissues, such as the skin, lungs, ovary, and so on, including secretory or glandular epithelia, such as mammary glands and the liver.

*Remark 3.2 (The Complexity of Cancer)*  Cancer is a set of complex chronic degenerative diseases, and may be considered as a robust manifestation of underlying developmental systems-level mechanisms at play during normal processes. But cancer is tightly associated with human aging, and at the same time, lifestyle factors seem to be modulators of the onset and progression of the disease. To understand such a complex disease it is important to uncover the intracellular regulatory mechanisms involved, and how they feed back to the cellular microenvironment, which, in turn, is affected by environmental factors and the individual's lifestyle. The modulation of the cellular and morphogenetic transitions that underlie the emergence and progression of neoplasias implies understanding the feedback among these intra- and extracellular mechanisms (see for instance [192] and [293] and the references therein).

## Emergence of Cancer from Complex Modulation of Regulatory Dynamics

The idea behind the project reported here is if in the progression of cancer we could identify conserved or generic patterns that result from conserved underlying systems-level mechanisms or core regulatory modules (see the previous chapter of this volume and [19] for a review of a similar approach used in the context of the study of cell differentiation and morphogenesis in plant systems). If we focus on epithelial cancer (i.e., carcinomas), we can indeed show that the clinical and pathological progression of cancer that originates from this type of tissue follows an almost generic pattern of cell transitions among patients with diverse types of carcinomas and of equally diverse origin, genetic background, or life condition. According to pathological studies, such pattern implies an intermediate pre-clinical condition in which hyperplastic tissues are established due to chronic inflammation (see for instance [254]). The cells implied in such tissues seem also to develop premature senescent signs and express genes that have been associated with inflammatory reactions and cell aging [260]. We hypothesized that such a repetitive pattern could emerge from the modulation of regulatory modules that are involved in normal cell function and in the organization of epithelial tissue. Such modules comprise multiple components and nonlinear feedback-based interactions among them, and are multistable, which is in fact a required dynamical characteristic as far as cell phenotypic plasticity is concerned.

*Remark 3.3 (From a Normal Epithelial State to an Anomalous Mesenchymal Phenotype)* Under normal conditions the regulatory modules that coordinate cell function and epithelial tissue organization display a specific configuration that leads to the maintenance of a normal epithelial cell differentiation state. But under certain tissue and microenvironmental conditions such systems transit to senescent states and from these to a mesenchymal stem cell-like state.

## A Core Regulatory Module Involved in the Onset and Progression of Epithelial Cancer

In what follows we focus on the core regulatory module that could be involved in the systemic mechanisms underlying the cellular and tissue transitions that eventually may lead to advanced carcinomas (see the details in [316]). With the model presented here we address why epithelial cancer seems to be highly reproducible among patients, and why it seems an almost inevitable and robust outcome of aging. Very importantly, based on stochastic models of complex gene regulatory networks we can also address what may be the modulatory role of random or stochastic fluctuations that recently have called the attention of cancer experts and that together with environmental and genetic factors, as well as epigenetic factors, lead to cancer emergence and progression. In contrast to targeting individual or a few molecules for prevention and treatment, our systems-level approach calls for preventive recommendations that may delay cancer emergence and/or slow its progression considering the systems-level mechanisms uncovered by using gene regulatory network modeling. We must point out that other researchers have also proposed similar developmental and network-based modeling approaches to cancer (see for instance [114, 212, 213, 262]).

We recently published a dynamic gene regulatory network model that integrates key molecular components involved in cell aging, cell cycle, metabolism, epithelial and mesenchymal differentiation, and inflammation [316]. All of these are processes that potentially underlie in vivo carcinogenesis and the cellular senescence of human epithelial cells, as well as their subsequent epithelial-to-mesenchymal transition induced by inflammation [298, 316].

Before going into the details of the gene regulatory network analyzed here, we briefly contrast in what follows two of the main paradigms in cancer biology: the classic genetic paradigm and the developmental perspective.

## Cancer: Nurture Versus Nature?

Traditionally, cancer is usually defined as a genetic disorder (see for instance [428]):

A diverse group of diseases that result as a consequence of changes at the DNA level.

The genetic view of cancer has a long history. Indeed, its lineage can be traced back to a series of fundamental discoveries. The causal link between genetic alterations (abnormalities of hereditary material) and cancer dates back more than a century, when chromosome aberrations were first observed in dividing cancer cells (see for instance [58, 193]). Indirect empirical support was subsequently provided by findings demonstrating that chemical damage to DNA causes both genetic mutations and cancer (reviewed in [285]). Moreover, DNA sequences obtained from cancers of diverse origin were shown to induce malignant transformation when introduced into human cell lines (see for instance [415, 416]). Similar observations further cemented a purely genetic causal view, having perhaps its maximal expression in the consequential mass media-based popularization of the oncogene and the tumor-suppressing gene concepts (see for instance [155, 259, 484]). Because of a large body of experimental work replicating such observations (see for example [186, 204, 269, 471]), it seems reasonable that nowadays most interpretations of cancer are subject to testing for consistency with a genetic origin. However, some recent cancer research breakthroughs have evidenced that in many cases genetic alterations are not sufficient to explain the oncogenic process, and that (micro)environmental factors such as chronic inflammation need to be taken into consideration.

## Random Mutations Do Not Explain Cancer

Because of this widely accepted belief in the genetic causes of cancer, the role of chronic inflammation and aging in the oncogenic process (see for instance [20, 38] is conventionally explained from the genetic view of cancer: it is generally assumed that aging and inflammation increase the chance of accumulating somatic mutations that constitute the ultimate cause of cancer. Genetic alterations, in turn, constitute the main source for the production of genetic instability that ultimately leads to cancer.

Notwithstanding the empirical evidence described above, some of the difficulties associated with an increasing number of inconsistencies derived from multiple reports of apparently causal genetic elements have been pointed out and critically reviewed several times [216, 424]. For example, it is not clear how a myriad of different types of random mutations affecting many different signaling pathways converge to a robust pathological phenotype, displaying the famous "hallmarks of cancer", including [192]:

- sustained proliferation
- dysregulation of cellular energetics
- resistance to apoptosis

The existence of a robust pathological phenotype calls for a systems-level exploration of cancer.

## Cancer as a Developmental Disease: Transcending the Classic Genetic Paradigm

An alternative view of cancer, originally proposed more than one and a half centuries ago (see for instance [470]), and that is increasingly receiving renewed attention, is that cancer can be considered as a possible pathological outcome of the same mechanisms underlying normal developmental processes [90, 216, 393, 470]. This developmental view of cancer aims to offer mechanistic explanations different clinical phenomena that cannot be understood under the classical genetic paradigm introduced above. For example, cancer cells have been shown to transition to morphological and transcriptional convergent phenotypes or genetic configurations irrespective of the tissue of origin [51]. Cancer behavior in some cells can be observed in the absence of mutations through trans- or dedifferentiation processes (see for instance [46, 274, 298, 499]). In addition, the malignant phenotype of cancer cells has been shown to be reversible ('normalized') by several experimental means and conditions not involving genetic modifications [294, 477, 490]. Observations such as these, and the fact that carcinogenesis invariably recapitulates processes normally occurring during embryogenesis [322], align with a developmental and multistable systems-level mechanism, rather than an entirely reductionist and purely genetic basis to understand cancer.

The developmental perspective has motivated our research group to propose a systems-level modeling framework based on the construction of integrative gene regulatory networks, as well as other types of models, that allows us to analyze cancer as a developmental process emerging from the same mechanisms underlying normal cell differentiation and morphogenesis.

*Remark 3.4 (Developmental Mechanisms and Genetic Mutations)* The developmental perspective of epithelial cancer does not cancel the possibility that certain genetic mutations, once they appear, may be critical to facilitate the transition to cancerous cellular behavior. In fact, integrative gene regulatory models as the one discussed here can re-conciliate the genetic and developmental paradigms of cancer, by offering at the same time a systems-level mechanistic explanation of the developmental dynamics underlying carcinogenesis (detailed below), and by providing a functional context for the interpretation of empirically observed genetic variants (see, for example [19, 83]).

In what follows, we first briefly discuss some of the methodological approaches currently in use in the cancer scientific research field, as well as novel proposals to interpret recent experimental findings. Then we introduce and discuss interpretations of some of the models proposed so far, and offer some perspectives on cancer prevention via modulation of lifestyle.

## *A Bottom-Up Developmental Perspective for the Understanding of Cancer*

As a methodological consequence of the intersection between the genetic tradition of cancer and the technological explosion in bioinformatics and biomedicine, massive resources are being devoted to genome sequencing of human tumors with the hope of finding underlying genetic causes and therapeutic targets (see for instance [164, 219, 486]). Unfortunately, despite the growing number of associative and descriptive analyses of cancer genomic data sets, very little is still understood about the dynamical mechanisms underlying the emergence and progression of cancer (see for instance [501]). Such understanding is mandatory for more rational preventive and therapeutic approaches. It is becoming clear that bioinformatics insights derived from genomic data analyses are not enough to predict or clarify the phenotypic pathological manifestations of cancer, and less so understand the systems-level underlying mechanisms. In an attempt to overcome this, major efforts are turning towards mapping epigenomic profiles (see for instance [388]). The cell and tissue specificity of the epigenome can then be used to contextualize and predict the potential disruptive role of mutations [253, 375].

It has been pointed out recently that, rather than using additional massive tumor genomic data, a better strategy for approaching cancer may be to develop methods for analyzing the molecular regulatory networks that underlie cancer emergence, progression, and therapeutic resistance at a systems-level (see for instance [102, 293, 501]). In line with both network-based approaches, a research initiative based on the interpretation of such genome-level data within the context of large-scale molecular networks following a top-down systems biology approach has been proposed recently [252, 316]. But still, the main goal of this approach is the molecular mechanistic conceptualization of potential cancer driver mutations (see for instance [207]).

Concerned with the current situation of cancer biology (i.e., the abundance of sequencing data, yet lack of mechanistic understanding), we and others have recently put forward a methodological viewpoint focusing on the systems dynamics of molecular networks following a bottom-up mechanistic approach aimed at understanding the generic patterns of cancerous stages [212, 217, 238, 316], instead of a descriptive data-based one (for discussions on systems biology modeling approaches, see [19, 112]). The goal of our ongoing effort is to provide conceptual clarity by means of generic models focusing on epithelial carcinogenesis. For this, we have started to integrate gene regulatory networks grounded on molecular experimental data. These systems biology approaches are inspired in our efforts to use plant systems to understanding cell differentiation and morphogenesis [19]. Also Stuart Kauffman (see for instance [238]) and Sui Huang and collaborators [214, 217], have made important contributions such as the celebrated *cancer attractor theory*, while others have proposed the *endogenous molecular cellular network hypothesis* [481, 510, 522]. The latter are much in line with the proposal that we pursue in this section.

The growing interest in modeling cancer development indicates an important research transition driven by multidisciplinary attempts to go beyond extensive, high-resolution description, towards uncovering and understanding systems-level mechanisms underlying generic patterns in the emergence and progression of cancer. This new perspective transcends the classical genetic paradigm, and its ultimate goal is to further our understanding and, hopefully, provide rational preventive and therapeutic strategies (fundamentally based on the modulation of the lifestyle).

In what follows we shall tackle epithelial cancer modeling via the analysis of how gene regulatory dynamics shape the epigenetic landscape, giving rise to the state-space trajectories that characterize the dynamics of epithelial cancer.

## *Gene Regulatory Networks and Epigenetic Landscape Modeling: The Case of Epithelial Cancer*

Understanding the emergence of phenotypic manifestations that characterize both health and disease requires integrative approaches based on the exploration of biological development and going beyond gene-centric studies (see for instance [19, 109, 112] and the references therein). Such system biology approaches have proven to be more powerful to propose predictive models, which is something that we are aiming for in the context of medical systems biology. In the last two decades, work in our research group has followed such an approach in the study of diverse systems-level developmental processes, specially using plant systems (see, for example [17, 19, 29, 50, 141]). Recently, we started adopting this approach to study epithelial carcinogenesis [316], work that we continue developing and extending. The rationale of our work follows the dynamical systems view of cell biology [112].

As reviewed above, a mechanistic systems biology approach to cell differentiation and morphogenesis relies on multistable and multi-level models with gene regulatory networks at its basis. As we previously pointed out, these are intracellular complex and highly nonlinear systems that comprise the underlying mutual gene regulatory interactions implied in developmental processes. Nonlinear dynamical gene regulatory networks yield several stationary states where the regulatory constraints imposed by the network are satisfied in a way that the expression of each gene stays unchanged (for a more in-depth explanation, see [19, 112] and the references therein). This general model has been proposed as a mechanistic explanation of how the same genome and network robustly generates multiple discrete cellular phenotypes during development (as discussed in [16, 19, 109, 216, 238, 318]). As in any nonlinear dynamical system, the stable stationary states are called attractors (recall the exposition turning around these concepts that presented in the previous chapters), and these states operationally correspond to configurations of gene or protein activation that underlie or correlate with different cell types or cellular phenotypes under study (which provides a systems-level explanation for phenotypic plasticity).

Gene regulatory network modules comprise sets of necessary and sufficient components and interactions to recover the stable configurations of activation under study. These configurations correlate with those that have been experimentally described for cell types or behaviors under study and comprise the expected attractors. Such modeling approach has been validated for several systems during flower development [18, 141, 318], stem cell differentiation [29, 273], and cell-fate decision [518], among many others in plants and animals.

## Epithelial Cancer Cellular Progression

We summarize here our work in epithelial cancer. This research work follows the modeling philosophy cultivated when studying developmental dynamics associated to cell differentiation and morphogenesis of plant structures. We focus on a core gene regulatory network module underlying the conserved time-ordered observed patterns of cellular transitions in the so-called carcinomas:

- Epithelial cells.
- Senescent cells.
- Mesenchymal-like cells.

We recently published an in silico model of the key cellular processes towards this cellular progression in vitro [316]. This corresponds to the spontaneous immortalization of epithelial cell lines. In vivo pathological studies have revealed a conserved pattern in the cellular transitions observed during the emergence and progression of this type of cancer. Normal epithelial cells transit to senescent ones after chronic inflammation. It is from these prematurely senescent cells that mesenchymal malignant ones emerge.

## *Assembling the Regulatory Network Grounded on Experiments Data*

We aimed at integrating a gene regulatory network module integrating the main molecular genetic processes involved in the emergence and transition of epithelial cancer detailed above (see for instance [298, 504]). It has been documented experimentally that epithelial cells that are exposed in vitro to cytokines undergo epithelial-to-mesenchymal-transition and the resulting cells manifest mesenchymal stem-like characteristics and genetic profiles as well as behaviors (e.g., capacity to migrate) (see Fig. 3.2). The resulting mesenchymal stem-like cells are very similar to cancer stem cells in vivo. Such cells have also been shown to have the potential to initiate cancer in murine models (see for instance [46, 272, 298, 329, 499]). The mathematical model comprising a dynamical mechanistic explanation for such epithelial-to-mesenchymal transition via a *senescent cell state* has been recently published [316]. In this study we used the gene regulatory network modeling approaches that have been described in this volume, especially those related with the postulation of multistable complex intracellular regulatory networks, to describe different cellular phenotypes underlying normal and altered transitions and tissues. Molecular components and interactions involved in:

- Cell cycle.
- Epidermal and mesenchymal cell differentiation.
- Senescence.
- Inflammation.
   and:
- Epigenetic silencing,

were incorporated into the large regulatory network reproduced in Fig. 3.3 (for details on the experimental evidence used to integrate and assemble this network please refer to the original publication [316]). All of the components incorporated have been experimentally characterized in normal development of epidermal and mesenchymal tissues, and the other processes including their involvement in cancer of various types with emphasis on carcinomas. Nonetheless, to test that we had incorporated components that are relevant to study cancer emergence, we pursued an network-based gene set enrichment analysis of the proposed gene regulatory network.

## *Network-Based Gene Set Enrichment Analysis*

To further support that the set of regulatory interactions that we manually curated based on published data are indeed representative of the cellular-level processes underlying epithelial carcinogenesis, we performed a network-based Gene Set Enrichment Analysis (GSEA; [430]) of the gene regulatory network, using both the KEGG and the GO Biological Process databases as reference [166, 231]. As a

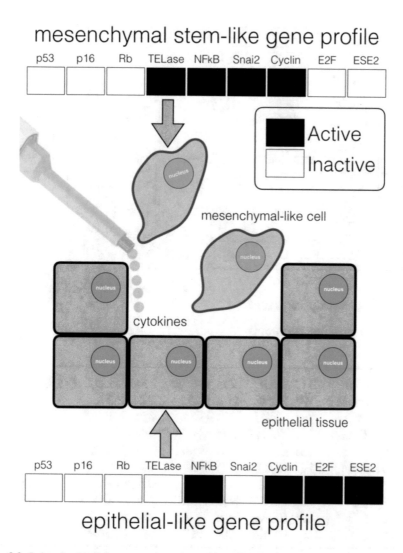

**Fig. 3.2** Induced epithelial-to-mesenchymal transition. This figure shows in a schematic manner that (under some conditions) cytokines and other microenvironmental conditions characterizing chronic inflammation can induce epithelial-to-mesenchymal transition, giving rise to mesenchymal-like cells from epithelial cells. Both the epithelial-like cells and the mesenchymal-like cells are characterized by a specific gene profile defined by the activity of well-characterized transcription factors (see for instance [316, 383])

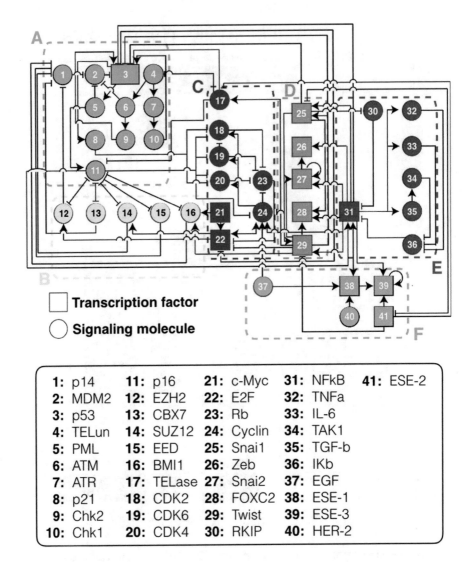

**Fig. 3.3** Gene regulatory network for epithelial carcinogenesis. Figure taken from [316], published under the Creative Commons Attribution 4.0 International License (http://creativecommons.org/licenses/by/4.0/)

result of this analysis, we found that among the 12 pathways or processes reported as significant when taking the KEGG database as a reference, 10 (≈83%) correspond to the cancer types:

- Bladder cancer.
- Chronic myeloid leukemia.
- Non-small cell lung cancer.
- Glioma.
- Melanoma.
- Pancreatic cancer.
- Prostate cancer.
- Small cell lung cancer.
  and:
- Thyroid cancer.

From these cancer types, six (66.6%) correspond to carcinomas. When taking the GO Biological Process database as reference, we found that the molecules considered in our regulatory network are significantly enriched for several of the biological processes known to play important roles during spontaneous immortalization of epithelial cells, including (see Table 1 in [316]):

- Replicative senescence.
- Cellular senescence.
- Cell aging.
- Positive regulation of epithelial to mesenchymal transition determination of adult life span.

Additionally to this GSEA, we performed a network-based topological gene set enrichment analysis (see Methods in [316] for the details) and found that, in addition to the enrichment of the pathways and processes described above, the molecules in the proposed network show also a topological signature that strongly resembles the structure of the cancer pathways included in the KEGG database.

## A Core Regulatory Network Module from the Reduction of the Original Network

To mathematically analyze the functional consequences of the proposed network, we simplified the large network shown in Fig. 3.3 into a smaller, computationally tractable Core Regulatory Network, by collapsing the linear pathways (i.e., pathways that do not involve feedback-based interactions).

For this, we applied a systematic knowledge-based reduction algorithm that reduces these pathways but preserves the regulatory interactions, obtaining the core regulatory module shown in Fig. 3.4.

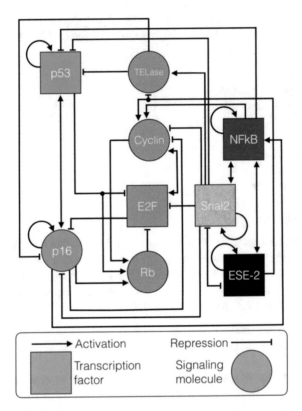

**Fig. 3.4** Epithelial-to-mesenchymal transition core regulatory network module. Core regulatory network module that underlies epithelial-to-mesenchymal transition in the context of epithelial cancer, developed in [316]. The nodes in blue are involved in senescence dynamics and the nodes in green characterize cell-cycle dynamics. When over-activated, the node in red (i.e., NF$\kappa$B) represents the inflammatory response shaping the transition dynamics. The nodes in orange and in black (i.e., Snai2 and ESE-2) represent the mesenchymal stem-like and the epithelial-like gene profiles, respectively. Figure adapted from [316], published under the Creative Commons Attribution 4.0 International License (http://creativecommons.org/licenses/by/4.0/)

This network module retained all the functional feedback motifs in the large network, but was small enough to be analyzed in terms of the dynamical behaviors and the epigenetic landscape modeling approaches described above [110]. Specifically, we aimed to test if the recovered core regulatory module contained a set of interactions that were necessary and sufficient to robustly converge to stable configurations or attractor states with the patterns of gene activation for the included components, that have been described in normal epithelial, senescent, and mesenchymal stem-like cells. Our analyses confirmed these hypotheses. Indeed, the gene regulatory core module (characterized by the logical rules shown in Box 3.1) attained only three attractors with activation patterns recovering the behaviors that coincide with experimental observations (see Fig. 3.5). We also found that these

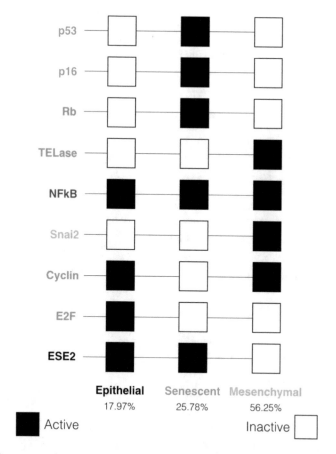

**Fig. 3.5** Attractors of the core regulatory network module underlying the epithelial-to-mesenchymal transition. The three attractors (steady-state configurations) of the core regulatory network module underlying epithelial-to-mesenchymal transition in the context of epithelial cancer [316]: the epithelial-like phenotype, the senescent phenotype, and the mesenchymal-like phenotype. Each attractor is displayed (from left to right) as a column vector that contains the set of binary variables that represent the activation states of the nodes of the core regulatory network module (i.e., the involved transcription factors and signaling molecules). The percentages represent the relative size of the different basins of attraction. Figure adapted from [316], published under the Creative Commons Attribution 4.0 International License (http://creativecommons.org/licenses/by/4.0/)

systems states were quite robust to different types of perturbations of the logical rules (see the details in [316]).

*Remark 3.5 (From Interactions to Attractors)* When exploring (through computer-based simulations) the epithelial-to-mesenchymal transition core regulatory network module shown in Fig. 3.4, we get the set of attractors shown in Fig. 3.5. These results are consistent with the available empirical evidence. When the attractors are known a priori, it is possible to build a network that converges to those attractors. However, in our context, we go from the characterization of regulatory interactions to resulting attractors, since we are interested in the uncovering of the circumstances of emergent stability.

---

**Box 3.1.** Epithelial-to-mesenchymal transition core regulatory module

This box includes the set of logical rules that defines the core regulatory module underlying epithelial-to-mesenchymal transition in the context of the onset and progression of epithelial cancer [316].

The nine logical rules constitute a discrete-time and discrete-space Boolean transcriptional gene regulatory network.

For each rule, the right-hand side of the rule defines the regulatory function that updates the activation state of the corresponding gene at updating time. Symbols $\wedge$, $\vee$, and $\neg$, stand for the AND, OR, and NOT logical operators.

**Nodes of the network**

p53 : Transcription factor, regulates cellular responses.
to DNA damage

p16 : Signaling molecule, inhibits cyclin.
dependent kinases

Rb : Signaling molecule, inhibits cell-cycle progression.

TELase : Signaling molecule, RNA-dependent DNA polymerase
that synthesizes telomeric DNA sequences.

NF$\kappa$B : Transcription factor, regulates the immune
response to infection.

Snai2 : Transcription factor, repressor of E-cadherin transcription.

Cyclin : Signaling molecule, regulates the progression
of cells through the cell cycle
by activating cyclin-dependent kinase (Cdk) enzymes.

E2F : Transcription factor, regulates genes required for
appropriate progression through the cell cycle.

ESE2 : Transcription factor, regulates late-stage differentiation
of keratinocytes (as well as glandular epithelia).

---

(continued)

**Box 3.1.**   (continued)

**Logic functions**

**Mesenchymal phenotype**

$\textbf{Snai2} = (\neg ESE2 \wedge \neg NF\kappa B \wedge \neg Snai2) \vee (\neg ESE2 \wedge \neg NF\kappa B \wedge Snai2)$
$\vee (\neg ESE2 \wedge NF\kappa B \wedge \neg Snai2) \vee (\neg ESE2 \wedge NF\kappa B \wedge Snai2)$
$\vee (ESE2 \wedge NF\kappa B \wedge Snai2)$

**Epithelial phenotype**

$\textbf{ESE2} = (\neg NF\kappa B \wedge \neg Snai2 \wedge \neg ESE2) \vee (\neg NF\kappa B \wedge \neg Snai2 \wedge ESE2)$
$\vee (\neg NF\kappa B \wedge Snai2 \wedge ESE2) \vee (NF\kappa B \wedge \neg Snai2 \wedge \neg ESE2)$
$\vee (NF\kappa B \wedge \neg Snai2 \wedge ESE2)$

**Cellular inflammation**

$$\textbf{NF}\kappa\textbf{B} = \neg (\neg ESE2 \wedge \neg p16 \wedge \neg Snai2 \wedge \neg NF\kappa B)$$

**Cellular senescence**

$\textbf{p16} = (\neg p16 \wedge \neg E2F \wedge p53 \wedge \neg TELasa \wedge \neg Snai2)$
$\vee (\neg p16 \wedge \neg E2F \wedge p53 \wedge \neg TELasa \wedge Snai2)$
$\vee (\neg p16 \wedge \neg E2F \wedge p53 \wedge TELasa \wedge \neg Snai2)$
$\vee (\neg p16 \wedge E2F \wedge p53 \wedge \neg TELasa \wedge \neg Snai2)$
$\vee (\neg p16 \wedge E2F \wedge p53 \wedge \neg TELasa \wedge Snai2)$
$\vee (\neg p16 \wedge E2F \wedge p53 \wedge TELasa \wedge \neg Snai2)$
$\vee (p16 \wedge \neg E2F \wedge \neg p53 \wedge \neg TELasa \wedge \neg Snai2)$
$\vee (p16 \wedge \neg E2F \wedge p53 \wedge \neg TELasa \wedge \neg Snai2)$
$\vee (p16 \wedge \neg E2F \wedge p53 \wedge \neg TELasa \wedge Snai2)$
$\vee (p16 \wedge \neg E2F \wedge p53 \wedge TELasa \wedge \neg Snai2)$
$\vee (p16 \wedge E2F \wedge \neg p53 \wedge \neg TELasa \wedge \neg Snai2)$
$\vee (p16 \wedge E2F \wedge \neg p53 \wedge \neg TELasa \wedge Snai2)$
$\vee (p16 \wedge E2F \wedge \neg p53 \wedge TELasa \wedge \neg Snai2)$
$\vee (p16 \wedge E2F \wedge \neg p53 \wedge TELasa \wedge Snai2)$
$\vee (p16 \wedge E2F \wedge p53 \wedge \neg TELasa \wedge \neg Snai2)$
$\vee (p16 \wedge E2F \wedge p53 \wedge \neg TELasa \wedge Snai2)$
$\vee (p16 \wedge E2F \wedge p53 \wedge TELasa \wedge \neg Snai2)$
$\vee (p16 \wedge E2F \wedge p53 \wedge TELasa \wedge Snai2)$
$\vee (p16 \wedge \neg E2F \wedge \neg p53 \wedge TELasa \wedge \neg Snai2)$

---

**Box 3.1.**   (continued)

$$\begin{aligned}
\textbf{p53} = &(\neg p53 \wedge \neg NF\kappa B \wedge \neg TELasa \wedge \neg p16 \wedge \neg Snai2) \\
&\vee (p5 \wedge \neg NF\kappa B \wedge \neg TELasa \wedge p16 \wedge \neg Snai2) \\
&\vee (\neg p53 \wedge NF\kappa B \wedge \neg TELasa \wedge p16 \wedge \neg Snai2) \\
&\vee (p53 \wedge \neg NF\kappa B \wedge \neg TELasa \wedge \neg p16 \wedge \neg Snai2) \\
&\vee (p53 \wedge \neg NF\kappa B \wedge \neg TELasa \wedge p16 \wedge \neg Snai2) \\
&\vee (p53 \wedge NF\kappa B \wedge \neg TELasa \wedge p16 \wedge \neg Snai2)
\end{aligned}$$

**Cell cycle**

$$\begin{aligned}
\textbf{Cyclin} = &(\neg ESE2 \wedge \neg E2F \wedge \neg p16 \wedge \neg NF\kappa B \wedge \neg Snai2) \\
&\vee (\neg ESE2 \wedge \neg E2F \wedge \neg p16 \wedge NF\kappa B \wedge \neg Snai2) \\
&\vee (\neg ESE2 \wedge \neg E2F \wedge \neg p16 \wedge NF\kappa B \wedge Snai2) \\
&\vee (\neg ESE2 \wedge E2F \wedge \neg p16 \wedge \neg NF\kappa B \wedge \neg Snai2) \\
&\vee (\neg ESE2 \wedge E2F \wedge \neg p16 \wedge NF\kappa B \wedge \neg Snai2) \\
&\vee (\neg ESE2 \wedge E2F \wedge \neg p16 \wedge NF\kappa B \wedge Snai2) \\
&\vee (ESE2 \wedge \neg E2F \wedge \neg p16 \wedge \neg NF\kappa B \wedge \neg Snai2) \\
&\vee (ESE2 \wedge \neg E2F \wedge \neg p16 \wedge NF\kappa B \wedge \neg Snai2) \\
&\vee (ESE2 \wedge E2F \wedge \neg p16 \wedge \neg NF\kappa B \wedge \neg Snai2) \\
&\vee (ESE2 \wedge E2F \wedge \neg p16 \wedge NF\kappa B \wedge \neg Snai2)
\end{aligned}$$

$$\textbf{TELasa} = (\neg Snai2 \wedge \neg ESE2) \vee (Snai2 \wedge \neg ESE2)$$

$$\begin{aligned}
\textbf{Rb} = &(\neg Cyclin \wedge \neg p16 \wedge p53) \vee (\neg Cyclin \wedge p16 \wedge \neg p53) \\
&\vee (\neg Cyclin \wedge p16 \wedge p53) \vee (Cyclin \wedge \neg p16 \wedge p53) \\
&\vee (Cyclin \wedge p16 \wedge \neg p53) \vee (Cyclin \wedge p16 \wedge p53)
\end{aligned}$$

$$\begin{aligned}
\textbf{E2F} = &(\neg Rb \wedge \neg p53 \wedge \neg Snai2 \wedge \neg Cyclin) \\
&\vee (\neg Rb \wedge \neg p53 \wedge \neg Snai2 \wedge Cyclin)
\end{aligned}$$

---

We also used the model to simulate several different genetic alterations, recovering the experimentally characterized expression profiles that have been characterized for several loss and gain-of-function mutants (see Fig. 3.6). With this agreement between model behavior and experimental data, we validated the gene regulatory core model. Further, this result exemplifies how a dynamical model can be used to systematically evaluate the robustness of a gene regulatory network to genetic perturbations.

Additionally, our model suggests a systems-level dynamical explanation to the fact that in many cases intermediate inflammatory cells are observed before cells transit to a mesenchymal state, because the epigenetic landscape modeling of the network under analysis recovered the observed time-ordered pattern (epithelial-

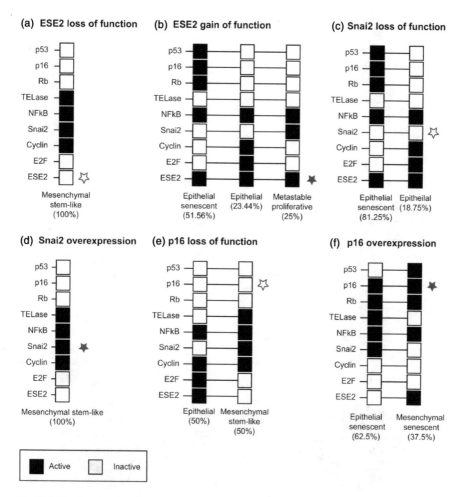

**Fig. 3.6** Predicted attractors of loss- and gain-of-function mutants of the GRN. Predicted attractors of loss- and gain-of-function mutants of the GRN for ESE2 (**a, b**), Snai2 (**c, d**) and p16 (**e, f**). Percent (%) represents the size of the corresponding basin of attraction. Figure taken from [316], published under the Creative Commons Attribution 4.0 International License (http://creativecommons.org/licenses/by/4.0/)

senescent-mesenchymal cells) repeatedly observed during epithelial cancer progression (see Fig. 3.7).

Interestingly, many of the components of the regulatory core proposed in our study had been pointed out as important genes involved in epithelial (and other types of) cancer, but the topology and architecture of the regulatory network that we recently published had not been proposed before. We propose that it is the dynamics of this gene regulatory network module what underlies the transitions from normal epithelial cells to senescent ones, and finally to mesenchymal cells with stem-like traits that are characteristic of carcinomas and that appear in vitro during spontaneous immortalization of epithelial cells.

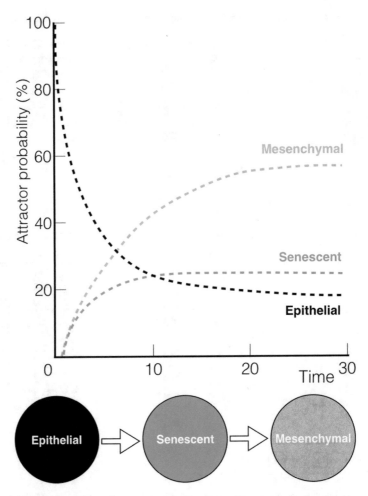

**Fig. 3.7** Temporal sequence and global order of cell-fate attainment pattern under the stochastic Boolean gene regulatory model during epithelial carcinogenesis. For the core gene regulatory network module underlying epithelial-to-mesenchymal transition, in the context of epithelial cancer, this figure shows the maximum probability of attaining each attractor as a function of time (in iteration steps). The most probable sequence of cell attainment is: epithelial → senescent → mesenchymal stem-like. Figure adapted from [316], published under the Creative Commons Attribution 4.0 International License (http://creativecommons.org/licenses/by/4.0/)

The results obtained for the regulatory module under discussion for epithelial cancer are in line with results obtained for other developmental systems. The regulatory processes involved in cell transitions during normal and cancerous development are prone to stochasticity as described above in the section devoted to epigenetic landscape modeling. Several years ago, we hypothesized that at least some aspects of the morphogenetic temporal and spatial patterns observed during normal and altered development emerge from the deterministic dynamical gene regulatory network underlying a particular process under study and its resonance with stochastic fluctuations that result from intrinsic and extrinsic noise [18]. We tested such stochastic explorations of the epigenetic landscape that emerges from floral-organ specification gene regulatory network and recovered the time-ordered patterns observed in floral development [18]. Interestingly, for the case of the core regulatory module involved in epithelial-to-mesenchymal transition, our stochastic simulations and exploration of the epigenetic landscape as succinctly summarized above, reproduced the experimentally observed time-ordered transitions of cellular phenotypes during carcinoma progression: initial epithelial cells more likely transit to senescent cellular states and then to mesenchymal stem-like cell states. These results strongly suggest that the uncovered regulatory core underlies important aspects of the cellular transitions that have been observed in vitro and in vivo, but also of their widely conserved behaviors or time-ordered patterns during spontaneous immortalization of epithelial cells in vitro and probably also in the progression of in vivo carcinomas. This model also supports that together with complex gene–gene interactions, noise is an important aspect of the emergence of cancer.

## Discussion on the Medical Systems Biology Consequences of the Exploration of the Model

As resulting from the exploration of the dynamical properties of the core gene regulatory network module underlying epithelial-to-mesenchymal transition, it is concluded that the transition depends on cell senescence. This is an important result. We must point out that Stuart Kauffman was among the first to postulate that disease-associated cellular states could correspond to particularly robust attractors that once attained were difficult to leave [238]. But experimental data to test such proposition has become available only in the last few years. More recently, several researchers (see for instance [87, 105, 217, 496]) have indeed supported this proposal and further developed it with gene regulatory network dynamical modeling approaches that can now be grounded on experimental data. Furthermore, there is evidence that shows that cancer progression can occur in the absence of genetic alterations sometimes, and some normally behaving cells might have some somatic mutations that are characteristic of some cancers, suggesting that purely gene-centric approaches for understanding cancer are in fact very limited. Our [316] and other models [87, 217] suggest that both the intracellular gene regulatory network implied in the cellular processes involved in cancer progression and its feedback with microenvironmental signals are involved in the dynamics of emergence and progression of cancer. Some of these models have also pointed out to the importance of stochasticity in cancer emergence and progression (see [15, 145, 151, 316]). Moreover, some authors have suggested that cancer results from pre-existent pathological attractors, which are not accessible under normal development but become accessible during illness (see [212, 216]). This view suggests that the attractors of cancer can be reached through the nonexclusive occurrence of perturbations to either the network state, by means of intra- or intercellular signals, or to the gene regulatory network structure by mutation (see [24, 211, 233]). Therefore, according to this theory: *cancer can be triggered when epithelial cells undergo abnormal state transitions towards attractors that encode embryonic phenotypes.* We propose here that perturbations can have re-structuring effects in the epigenetic landscape, such that previously unstable states become stable or even robust attractors. The latter view is more dynamical in nature and suggests that the epigenetic landscape can be reshaped due to non-genetic deterministic (chemical and physical fields or environmental factors) or stochastic dynamics, during which formerly unstable states become stable.

*Remark 3.6 (Epigenetic Landscape Re-shaping)* In a recent study our research team uncovered a case of epigenetic landscape re-shaping while modeling plant development in a de-differentiation case that may be similar to what occurs in some types of cancer. The over-expression of a MADS-domain transcriptional regulator generated a novel attractor that shared both differentiated and stem-cell gene expression profiles and thus explained the behavior of some cells in the flowers of such gain-of-function lines (see the details in [369]). Such studies to understand the systems-level mechanisms underlying phenotypic plasticity in plants can also provide insights into human development and the emergence of disease conditions. In any case, the key point of the type of proposal we put forward here is that the potential for manifesting a cancerous phenotype is intrinsic to the human genome and regulatory networks at play during normal development. It is perhaps an inevitable consequence of metazoan evolution (as has been pointed out in [212]).

On the other hand, the approach proposed here and in other papers (see [102, 213]) may be useful to reconcile both the genetic and developmental views of cancer, because developmental dynamics and the influence of environmental and microenvironmental factors result in the establishment of pathological attractors; these may imply altered proliferation dynamics, which, in turn, may promote higher rates of mutations, and in any case, genetic perturbations affecting gene regulatory network structure may facilitate that cells, specifically after chronic inflammation and premature senescence, attain new attractors that yield abnormal or pathological cell behaviors

The modeling work in cancer described here enables us to conclude that an "abnormal" cell state associated with a cancerous phenotype is readily attainable with high probability as a consequence of the developmental dynamics of the gene regulatory network that we have uncovered. This has two potential interpretations:

**Primo:** The architecture of the uncovered network could include structural alterations resulting from genetic perturbations that are not considered in this model explicitly. This first interpretation may be postulated given that most of the experimental work that we considered for the assembly of this model came from cancer-specific experimental data. If this is the case, in agreement with the cancer attractor theory, we could say that we have found a perturbed network whose architecture increases the likelihood of a transition from a normal epithelial to an abnormal pathological phenotype.

**Secondo:** It is possible that both the architecture and the likelihood of transitions recovered when using the model presented here do not imply genetic alterations, but the transition rates among cell types could be modulated by both genetic and non-genetic factors. This would imply an intrinsically mutation-free model mechanism for normal and abnormal developmental dynamics, and the latter could cause a disease condition when attained in an ectopic manner both in terms of temporal and spatial aspects of morphogenesis. In this case, the cancer state might correspond to a novel attractor that emerges in the epigenetic landscape as a consequence of the over-expression of one transcription factor, as we found in plants case [369]. It is also possible that it could result from a normal attractor that is visited during development, but when attained ectopically it implies a morphogenetic alteration referred to as cancer at the tissue level. Finally, it could imply a non-stable attractor, which, as a consequence of a non-genetic alteration of the epigenetic landscape, becomes stable and more cells attain such abnormal state. We favor this last interpretation. To further consider the latter, below we elaborate on the level of abstraction of the model presented here and the level of organization at which it is valid.

## *The Promotion of Inflammation by Senescent Cells Increases the Likelihood of Epithelial-to-Mesenchymal Transition*

Considering cellular senescence and inflammation, and distinguishing between in vitro and in vivo processes, is fundamental to clarify the interpretation of our model. We first discuss what occurs in actual human tissues and the importance of chronic inflammation and premature cellular senescence in the emergence of epithelial cancers. As senescent cells increase in number within normal tissues under normal aging or abnormal chronic inflammation, aged tissues are prone to have a pro-inflammatory milieu associated with immune system infiltration and the active secretion of pro-inflammatory molecules (e.g., cytokines) by senescent cells [96]. This causes structural damage of tissue that might also imply alterations in the physical fields that in turn impact the mechano-sensitive signal-transduction pathways. The latter are likely interconnected to the gene regulatory network uncovered in the model discussed in this book and require further study. Indeed, it has been described that the action of cytokines and associated inflammation also increases the probability of epithelial-to-mesenchymal transition [38]. All of these alterations are not necessarily directed to specific genetic mutations, although some of the latter might for sure make tissues more prone to inflammatory responses and premature cellular senescence. But this would not imply a direct causal relationship between a specific mutation and the emergence of a cellular altered state that could correspond to a cancerous one. Hence:

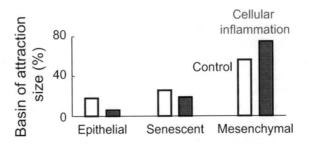

**Fig. 3.8** Chronic inflammation affects the sizes of the basins of attraction. Chronic inflammation affects the propensity to converge to a mesenchymal-like phenotype by altering the sizes of the basins of attraction. Figure taken from [316], published under the Creative Commons Attribution 4.0 International License (http://creativecommons.org/licenses/by/4.0/)

> We propose that chronic inflammation facilitates the transition from an epithelial to a senescent state and finally to a mesenchymal state, favoring the onset and progression of cancers emerging from epithelial tissues in vivo.

In terms of the model of epigenetic landscape described in previous chapters of this volume, the above behavior could imply that:

1. The landscape is altered by non-genetic factors mainly (although some genetic alterations might make such alteration more or less feasible, as shown in Fig. 3.6) and as a consequence, an attractor corresponding to the mesenchymal-state becomes available or more likely (see Fig. 3.8, showing the effects of chronic inflammation on the sizes of the basins of attraction) or a new de-differentiation attractor does.
2. Under normal conditions the mesenchymal state might not be easy to access once in a epithelial tissue, and consequently cancer is not so frequent. Hence, it might be quite distant from the epithelial one, but the senescent state, which the cells may attain due to chronic inflammation, might be closer to the mesenchymal state or makes this attractor more accessible in the epigenetic landscape. Alternatively, a non-stable state such as one that combines the expression characteristic of mesenchymal and stem cells becomes more stable and accessible, also once cells are in an inflammation state.

We are aware that in its present state the model used here to exemplify our approach to study the emergence of cancer only considers the intracellular minimal and sufficient set of restrictions to recover the three types of cells being considered and that have been observed in the great majority of epithelial cancers.

*Remark 3.7 (Multi-Level Dynamics)* Multi-level models of dynamics of cancer are required to further consider the tissue-level components and the feedback between the intracellular gene regulatory network module and the microenvironment, and the physical fields. These are likely to be critical modulators of the transition and of the time that cells take to go from a normal state to a cancer-like one.

Interestingly, in vitro cells spontaneously transit from an epithelial to a mesenchymal state. This fact alone validates our model and suggests that in contrast to cells in vivo, in the dish, cells are able to transit to a mesenchymal state spontaneously. Alternatively, intermediate states are also visited in vitro, but have not been characterized or further studied. Interestingly, the induction of epithelial-to-mesenchymal transition in immortalized cells has been experimentally shown to produce cells with both reduced expression of the senescence markers p16 and p53, and gain of the enzyme telomerase, both of which allow the cells to surpass senescence [162]. Phenotypically, the resultant cells:

1. Are similar to cancer stem cells, tumor-initiating cells, or embryonic stem cells [329].
2. Display resistance to apoptosis.
3. Have the ability to migrate, metastasize, and form secondary tumors—all lethal traits characterizing cancer cells [298].

Therefore, it seems that cells that have reached a senescent phenotype are prone to acquire stem-like properties under a pro-inflammatory environment. Coincidently, recent work has started to characterize molecular similarities between senescent and cancerous cells [104].

## Model-Based Interpretation Of Cancer Dynamics

The experimentally grounded gene regulatory network model presented here already recovers the dynamical behavior observed during the acquisition of stem-like properties by epithelial cells in vitro. However, the current model only considers intracellular dynamics, thus predicting cellular-level behavior. At this level, a generic series of cell-state transitions widely observed and robustly induced by inflammation in cell cultures seem to naturally result from the self-organized behavior emerging from the underlying regulatory network. Interestingly, similar processes are known to be instrumental during embryogenesis, a developmental stage lacking the adult tissue aspects highlighted above. In particular, senescence is also a natural process fundamental for early development in mammals (see [333, 427]), and epithelial-to-mesenchymal transition is known to have a critical role during embryogenesis (see for instance [298]). We reasoned that cells within an aged tissue might somehow be prone to revisit the developmental processes that originally shape the embryo and that spontaneously occur in vitro. If that is the case, then:

> What conditions within an aged tissue could trigger such path, and why is it so prevalent under certain conditions?

This is an important question. Let us explore an answer based on the proposed model.

An aged tissue is prone to suffer architectural deterioration and to present a pro-inflammatory environment; both aspects associated with the increase of senescent cells (see for instance [73]). We suspect that such tissue-level conditions, which are associated with a bad prognosis in cancer, may increase the rate of occurrence of the cell-state transitions observed in vitro due to the promotion of embryonic processes. Specifically, under such conditions senescent cells are likely to undergo epithelial-to-mesenchymal transition in vivo. In support of this, in addition to recovering the ordered state transitions, our model predicts that constitutive expression of the inflammatory pathway by the action of NF$\kappa$B increases the likelihood of acquiring a mesenchymal stem-like phenotype (see Fig. 3.8). This situation is likely to occur in vivo due to a feedback mechanism established by the secretion of inflammatory signals by senescent cells that reinforce local inflammation due to a consequential increase in immune infiltration.

*Remark 3.8 (Nutrition and Cancer: The Key Role of Inflammation)* As pointed out in [523]: (1) There is compelling evidence that nutrition has considerable effects on the incidence and progression of cancer and responses to treatment. (2) A lifestyle characterized by caloric excess, sedentarism, and a high-fat, high-sugar Western-style diet tends to promote carcinogenesis. The processes at play include but are not limited to: increased inflammatory reactions; diminished immunosurveillance; and a considerable abundance of energy-rich metabolites (or trophic factors).

Considering the results of our modeling efforts and the empirical evidence highlighted above, the interpretation that links cellular and tissue-level descriptions proposed here goes as follows:

1. At the cellular level, the time-ordered cell-state transitions undergone by an epithelial cell subject to replicative senescence and subsequent inflammation result in the establishment of a mesenchymal stem-like state that might be eventually responsible for the origin of carcinomas in vivo.
2. At the tissue-level, the accumulation of senescent cells, and the associated induction of a pro-inflammatory state, promote cell-state transitions and set the stage for the progression to a malignant phenotype by eliminating tissue restrictions.

Importantly, this latter condition can be intensified in tissues that are subject to high proliferation rate because of the lifestyle choices of the individual (e.g., the lack of exercise, smoking, pro-inflammatory diet, exposure to toxic agents, stress, etc.). Hence, the "abnormal" disease-associated character of the natural dynamical process uncovered by our model may rest on the fact that:

> Underlying systems-level developmental mechanisms are reused out of context due to extracellular perturbations that inevitably occur during aging or chronic abnormal inflammation. We refer mainly to the disruption of tissue-level self-organizational processes normally at play in a healthy adult organism.

We can at this level discuss preventive therapeutic interventions based on what the proposed model implies.

## *Lifestyle Choices: Setting the Stage for the Risk Modulation of Cancer*

An intuitive consequence of the cancer viewpoint put forward here is the likelihood of risk modulation. Let us see how the results derived from the analysis of the proposed model can inform the design of preventive strategies.

Considering the cellular mechanism proposed above (transition from the epithelial-like phenotype to the mesenchymal stem-like phenotype via an intermediary senescent phenotypic state), a decrease in either or both the rate of accumulation of senescent cells and inflammation is likely to have a retarding effect on the onset and progression of cancer. The latter would retard the onset of the first alterations and/or slower the rate of cell state transitions. How feasible is it to achieve such retardation effects in reality? It is very feasible, and diet provides in fact a lifestyle modulatory mechanism that can delay the onset and progression of cancer.

## *Modulating Transitions Through Nutrition (and Other Lifestyle Choices)*

Empirical evidence strongly supports the beneficial effects of caloric restriction, fasting regimes, and so-called functional foods, which ultimately lead to a significant increase in at the cellular and tissue levels these habits indeed seem to promote healthy cellular environments that could retard the onset of cancer, at least in model organisms (see for instance [287, 288, 367, 523]). Further, recent epidemiological studies suggest a potential role of diet in certain human cancers, an effect that may be driven or mediated by lifestyle factors (see for instance [182]). Risk factors (e.g., obesity and sedentarism) are intertwined, and this should be taken into consideration when considering the modulation of lifestyle as a preventive therapeutic strategy to retard the onset of cancer.

Although the beneficial effects of a healthy lifestyle and environment for disease is increasingly being acknowledged [287], our model suggests an underlying

molecular mechanism to further study their relevance. Abundant research of the
regulatory processes and epigenomic modification associated with a healthy diet in
normal and pathological conditions is certainly required, nonetheless. Irrespective of
genetic background, understanding how the modulation of the environment delays
chronic degenerative diseases seems to be a promising endeavor. It must be pointed
out that the interplay between obesity and chronic inflammation has been well established (see for instance [382, 463, 498]). This provides a potential process-based
explanation by why caloric restriction attenuates chronic inflammation, resulting
then in the attenuation of cell dynamical processes giving rise to carcinogenesis.
We consider that the epigenetic landscape formalism can be a useful framework to
tackle the interplay between lifestyle and cancer dynamics. The effects of preventive
mechanisms on the shape of the landscape can provide holistic systematic tools to
identify components and processes that can increase the robustness of the stability of
healthy attractors, and delay the emergence and progression of altered cellular states.
This offers a research agenda intended to fight cancer through preventive modulation
of lifestyle (taking diet into account as the main variable in the therapeutic approach
equation).

We can at this level conclude our exposition on the gene regulatory network
dynamics underlying epithelial-to-mesenchymal transition, discussing some perspectives.

## *Final Comments and Perspectives*

The bottom-up medical systems biology approach exemplified here with cancers
that originate from epithelial cells (using for this discrete-time and discrete-space
Boolean models) is related to similar theoretical/conceptual proposals (see [216,
504]) and models (see for instance [140, 476, 514]). The systems-level mechanistic
understanding of the cellular-level processes integrated in the model discussed here
constitutes a first step to unravel key processes that might be at play in vivo during
the emergence and progression of neoplasias associated to different environments
and genetic backgrounds as exemplified with cancers that originate in epithelia.
Testing modeling hypotheses and predictions awaits the side-by-side development
of multi-level models integrating tissue-level processes with both in vitro and in
vivo perturbation experiments. We believe that the cellular-level network model
discussed here [316] is, nonetheless, a valuable building block for more detailed
modeling efforts integrating further sources of tissue-level constraints such as:

- Cell cycle progression.
- Cell–cell interactions.
- Differential proliferation rates.
- Chemical fields.
- Mechanical forces.

Finally, we stress that, beyond theoretical arguments, the intuition gained by our simple gene regulatory network model has important clinical implications that question current therapies that rather than slowing or reverting epithelial-to-mesenchymal transition, tend to promote cellular senescence, and even stemness of malignant cells (see, for example [142, 274, 323]). The view put forward here, if correct, would suggest the need for alternative ways of treatment of cancer, and it also would support novel strategies to prevent or delay the onset and progression of epithelial cancer. Elucidating more realistic and experimentally grounded dynamic network attractor models will ultimately help overcome fundamental obstacles in the prevention and treatment of cancer (see for instance [102]). Rational treatment alternatives following such a view have been already discussed elsewhere (see for instance [217]). Nevertheless, we consider that prevention and modulation strategies based on controllable environmental factors are promising directions to tackle the contention of cancer and other age-related chronic degenerative disorders alike. Hopefully the present work and discussion will motivate novel approaches to think about cancer research, prevention, and treatment.

It is time now to tackle, as our second study case, the systems-level modeling of chronic inflammation.

## 3.3   Chronic Inflammation

### *Motivation*

In this section, we present a mathematical model intended to describe the differentiation of CD4+ T cells in response to different microenvironmental conditions that characterize healthy and disease states. Since CD4+ T cells orchestrate the adaptive immune response in vertebrates, the analysis of this model yields important insights in the establishment of aberrant immune responses that can eventually lead to chronic inflammation. Specifically, we show how the modeling methodology proposed in this book is applied to construct and characterize a minimal regulatory network for the core transcription factors and signaling pathways involved in the cell-fate attainment of CD4+ T cells. This example illustrates the regulatory role of the feedback-structured interplay between the intrinsic or intracellular regulatory core and the extrinsic microenvironment. Based on the exposed model, some possible therapeutic interventions intended to carry out the modulation of the immune response are proposed.

## *Stability, Plasticity, and the Immune System*

Organisms live in a changing environment. In some cases, organisms must ignore these changes and maintain a stable phenotype, while in others they must react to them. How to distinguish between these two situations is not trivial, as living beings must take into account their internal state and the environmental cues that surround them. This means that to survive organisms require phenotypes that are both stable and plastic, two seemingly contradictory phenotypic traits. Which mechanisms enable living beings to be robust is still an open question.

The immune system exemplifies how organisms require both stability and plasticity. The immune system defends the organism against a wide range of pathogens and immune challenges. To completely control a pathogen, the immune system must maintain a response until the pathogen has been cleared. Failure to mount an effective immune response leads to chronic infections. However, as both the immune response and the infection progress, the circumstances change. Once the immune challenge has been overcome, the immune system must regulate itself to avoid autoimmune diseases. In this way, maintaining a "healthy" state requires both a stable response to clear the pathogens and plasticity enough to adapt to the changing immune challenges (see for instance [339]).

## *A Modeling Framework to Understand Immunity Dynamics*

The key question is understanding how complex biochemical interactions maintain the fine balance between plasticity and robustness of CD4+ T cells in homeostatic conditions, and how perturbations affect this balance eventually leading to disease. Answering this requires an integrative, systems-level modeling framework, that encompasses the regulatory interplay between multiple types of molecules that collectively shape phenotype decisions. Further, the model must take into account that the immune response is dynamic and heavily influenced by the environment. Also, it must be able to recover:

- differentiation patterns,
- plasticity, and:
- robustness

of the system. Finally, the model should be understandable and make predictions that can be experimentally validated by both wet lab and in silico scientists.

As discussed in the previous chapter, the simplest modeling framework that satisfies these requirements is the one based on discrete Boolean networks (see for instance [240]). Discrete-time and discrete-space Boolean networks integrate the available information of the molecular regulation to predict cellular-level phenomena using a mathematical formalism. As discussed previously, these networks consist of nodes -that represent genes, proteins, or other biological processes- and edges- that represent the regulatory interactions among the nodes. As Boolean networks are dynamical systems, it is possible to construct logical functions that describe the state of the nodes depending on the state of its regulators through time. The value of the node represents whether the gene or protein is active or inactive in the biological system. As far as the effect of the environment is concerned, it can be included in these models as input nodes. These functions are then evaluated to obtain the attractors of the network, which represent cell types or biological processes like the cell cycle [240]. Furthermore, as illustrated with the previous example, Boolean networks let us simulate multiple types of perturbations (genetic and non-genetic), which makes them ideal for studying cell-fate attainment. These are the reasons why Boolean networks have been extensively used to study how the cellular phenotypes raise from the molecular regulation (see for instance [8, 39, 317, 320, 341]).

The dynamical behavior of CD4+ T cells, our current subject of study, is conditioned by a changing environment. Moreover, the internal regulation of these cells can be affected by [521]:

- Developmental noise;
- Mutations;
- Environmental fluctuations.

To understand the robustness of these cells it is necessary to study both their stability and plasticity in a global context. Furthermore, it is important to develop methods to quantify this robustness and to determine the key components of the system. Boolean models let us study robustness and verify our model against many types of available biological information. For example, as already discussed, it is possible to study the effect of loss and gain of function mutations, and of changes in the microenvironment [320] in the phenotypic convergence. These models can also be used to study the effect of transient perturbations in the intrinsic components of the network and the inputs of the system [303]. Boolean models can also be used to:

1. Check if there is over-fitting.
2. Check if errors in the construction of the functions will affect the results.
3. Predict missing regulatory interactions (see for instance [28]).

Let us now briefly describe the regulatory role of CD4+ T cells.

**CD4+ T Cells**

CD4+ T cells are part of the adaptive immune response and help coordinate the different mechanisms of the immune response. Each of the CD4+ T cell types activates or inhibits different branches of the immune response (see Fig. 3.9). Namely:

- Th1 cells are associated with the response against intracellular bacteria and protozoa.

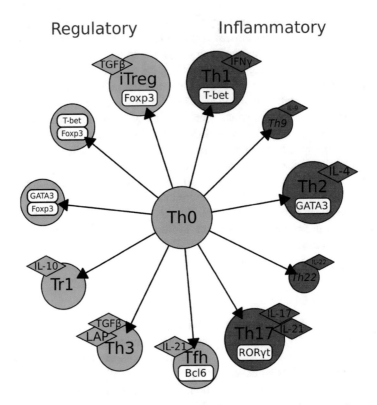

**Fig. 3.9** CD4+ T cell fate attainment. CD4+ T cell types are characterized by their unique cytokine production profiles, transcription factors, and biological functions. The main cell types are Th0, Th1, Th2, Th17, iTreg, and Tfh. Other cell types are IL-9 (Th9), IL-10+Foxp3-(Tr1), and TGF-$\beta$+Foxp3-(Th3) producing cells (see [303])

- Th2 cells are associated with the response against extracellular parasites including helminths [331].
- Th9 cells are associated with the response against parasites like *Trichuris muris* and *Nippostrongylus brasiliensis* [175, 237, 290, 406].
- Th17 cells are associated with the response against extracellular bacteria and fungi [467].
- Tfh cells are associated with the follicles and B cell maturation [62, 75, 103, 402, 469].

Furthermore, there exist multiple types of regulatory T cells, like Treg, Tr1, and Th3, that produce IL-10 and TGF$\beta$ and induce immune tolerance and control autoimmune diseases [165, 179, 268, 396, 485].

CD4+ T cells differentiate in response to the cytokines in their microenvironment. Cytokines can be produced by the same cell (intrinsic) or by other cells of the organism (extrinsic). Cytokines bind membrane receptors and activate signaling cascades that ultimately trigger the translocation of transcription factors to the nucleus

[357]. Complex signal integration occurs, partly due to the convergence of pathways in common nodes. For example, SOCS proteins compete with transcription factors (STATs) for the phosphorylation site in the receptor [158, 249, 506, 507]. Once the signal has arrived to the nucleus the transcription factors can activate or inhibit other transcription factors and cytokines, biasing the differentiation of the CD4+ T cell into different subsets [137, 234, 480]. This genetic control is also influenced by other factors like epigenetic marks and metabolism [137, 292]. The cytokines produced by the cell are secreted to the microenvironment, where they will join the cytokines produced by other cells of the immune system of the organism. The signals in the microenvironment are fundamental for cell-fate attainment and maintenance of these cells [56, 188, 280, 503].

CD4+ T cells have an heterogeneous transcriptional profile and can transdifferentiate in response to changes in the microenvironment [55, 127, 334, 358]. There is also considerable overlap among the expression profiles of different CD4+ T cells. There are reports of hybrid Treg/Th17, Treg/Th1, and even Th1/Th2 hybrid cells [248, 268, 500]. The regulatory cytokine IL-10 can be secreted by Th1, Th2, Th17, iTreg cells, and a variety of other immune cells [210, 399].

Once differentiated, CD4+ T cells can dynamically change their expression patterns as the immune challenge and the signals in the microenvironment change. These plastic transitions between cell types have been associated with maintaining the homeostasis of the organism and with some diseases. For example, the transition from Treg to Th17 has been associated with anti-tumor response, but also with multiple sclerosis and psoriasis [223]. There are restrictions to this plasticity; some transitions are more common, like the Treg/Th17 transition, while others seem to be uncommon, like the Th1/Th2 transition [55, 127, 223].

CD4+ T cells are closely integrated with the rest of the organism, and there is a strong relationship with the metabolism and the microbiome. For example, obesity-associated chronic inflammation (OACI) is characterized by a feedback loop between the inflammatory response of the immune system and the altered metabolism in obesity. In OACI there is an increase in inflammatory Th1 and Th17 cells, and a decrease in Tregs and IL-10 production. Hyperinsulinemia, which is associated with obesity and metabolic syndrome, inhibits IL-10 and decreases the number and stability of Treg cells [191]. At the same time, the species present in the gut microbiota can affect the differentiation of CD4+ T cells. Understanding the relationship between:

- the immune response,
- the metabolism,
- and the microbiome

is an open question.

*Remark 3.9 (Immune Response and Robustness)* Defining the phenotype of CD4+ T cells is not trivial, as we must take into account the heterogeneous transcriptional profiles, the dynamical response to the environment, the plastic transitions between cell types, and its relationship with the rest of the organism. This dynamical cellular behavior of CD4+ T cells is the result of a complex regulatory network of transcription factors, signaling pathways, and extracellular cytokines. Understanding how this regulatory network underlies cell-fate attainment and enables CD4+ T cells to maintain their function in the face of a changing environment can help us understand not only the immune response, but shed light on living beings achieve robustness.

## *Master Transcription Factors*

A first step for modeling CD4+ T cell-fate attainment is to construct a regulatory network from experimental data. For this, first focus on the transcriptional regulatory core that is formed by the interactions between master transcription factors (MTF). MTF are defined as the transcription factors whose expression is considered both necessary and sufficient to induce the differentiation of the cell towards a certain phenotype. To determine whether a minimal transcriptional regulatory core can explain the cell-fate attainment of CD4+ T cells, we proposed a Boolean model that consists of the regulatory interplay between the MTF for Th1 (T-bet), Th2 (GATA3), Th17 (ROR$\gamma$t), and Treg (Foxp3) [521]. The edges of the proposed network model correspond to the regulatory interactions between MTF (see Fig. 3.10a).

To construct this network, we made certain simplifications and assumptions. Specifically, although MTF collaborate with other transcription factors and are modulated by external signals transmitted by signaling pathways, these are ignored in this model. To do this, the cytokines and signaling pathways are modeled as input nodes. Using a bottom-up approach, the biological information used in this simple model has been obtained from multiple articles and curated databases to warrant its reliability. However, it must be pointed out that there exist computational algorithms to infer the network structure and the corresponding functions from transcriptomic data [116]. Using this experimental information it is possible to reconstruct both:

- The topology (as shown in Fig. 3.10a).
- The functions of the Boolean regulatory network (see Fig. 3.10c).

*Remark 3.10 (Network's Topology Is Not Enough)* The topology tells us which nodes of the system interact, while the functions describe the dynamics of this interactions. The topology alone is not sufficient to describe the dynamics of the system. For example, the topology alone cannot tell us if there is synergy, or not, between two regulators. This means that for most topologies there is more than one set of functions that can describe the dynamics of the system.

Constructing the dynamical functions allows us to formally represent synergistic interaction, distinguish between weak and strong inhibitions, and so on. Thus, this model formulation serves as a formal way to review and integrate the available experimental information, and helps us to determine which information is available and which areas require more research.

## (A)

| Source | Target | Interaction |
|--------|--------|-------------|
| Tbet | T-bet | Positive feedback |
| GATA3 | GATA3 | Positive feedback |
| Foxp3 | Foxp3 | Positive feedback |
| T-bet | GATA3 | Mutual inhibition |
| RORγt | Foxp3 | Mutual inhibition |
| T-bet | RORγt | Inhibition |
| GATA3 | RORγt | Inhibition |

Known
interactions

## (B)

Regulatory
graph

## (C)

$TBET_{t+1}$ = ($TH1s_t$ or $TBET_t$) and not $GATA3_t$

$GATA3_{t+1}$ = ($TH2s_t$ or $GATA3_t$) and not $TBET_t$

$RORGT_{t+1}$ = $TH17s_t$ and

not ($TBET_t$ or $GATA3_t$ or $FOXP3_t$)

$FOXP3_{t+1}$ = ($TREGs_t$ or $FOXP3_t$) and

not $RORGT_t$

Boolean
functions

## (D)

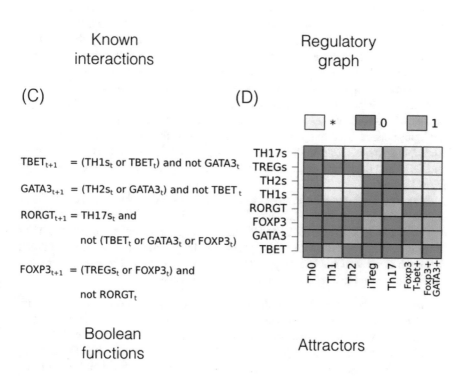

Attractors

**Fig. 3.10** Network of master transcription factors involved in CD4+ T cell-fate attainment. (**a**) Known interactions between master transcription factors based on published experimental data. (**b**) Graph of the CD4+ T cell regulatory network where nodes represent master transcription factors. Activations among elements are represented with black arrows and inhibitions with red blunt arrows. Dotted arrows represent inputs. (**c**) Boolean functions of the network. (**d**) Attractors of the network, arranged in columns. Each node can be active (green), inactive (red) or either (yellow).

By reconstructing the network, we can also observe some interesting patterns. For example, as shown in Fig. 3.10b, most MTF are self-regulated in a positive feedback manner. At the same time, most MTF have mutual inhibitions between them, creating negative feedback loops. Some of these patterns have been associated with specific biological regulatory functions (see for instance [325]).

In the network, each node represents a MTF (see Fig. 3.10b). The dynamical state of the node depends on the regulatory function (see Fig. 3.10c), and represents whether the MTF is present (1) or absent (0). The state of a specific node in the next time step $t + 1$ will depend in the states of its regulators during this time step $t$. We can evaluate the functions to determine the global state of the system in the next time step. The state of all nodes at a given time is the state of the system. We represent this as a string of 0s and 1s (see Fig. 3.10d, where each position corresponds to a given node).

One of the advantages of a Boolean network is that it lets us simulate an initial condition and follow the behavior of the system through time (as shown in Fig. 3.11). When, eventually, the system remains in a state, we say that we have found an attractor. All the transient states visited in the path to the attractor are part of the corresponding basin of attraction. As previously discussed, the attractors of the system correspond to cell types (i.e., specific cellular phenotypes). If we do this analysis for all possible states, we can determine all the attractors of the network (see Fig. 3.10d). If we suppose that there are no external signals (this means, that the input node is fixed to 0), the dynamic analysis of the network of MTF recovers attractors corresponding to different types of CD4+ T cells: Th0, Th1, Th2, iTreg, and Th17 and the hybrid states T-bet+Foxp3+ and GATA3+Foxp3+ (see Fig. 3.10b) [126, 127, 521].

However, the network converges to a configuration that characterizes the Th17 cells only in the presence of constant Th17 polarizing signals. This implies that the expression of ROR$\gamma$t, the Th17 MTR marker, is not sufficient to sustain the Th17 phenotype once the environmental triggers have been removed, requiring additional factors to generate a stable cell type. This result may be caused by the lack of feed-forward loops in the transcriptional regulatory network.

*Remark 3.11 (Why does* Th17 *Cells Dynamics Strongly Depends on Their Environment?)* Thanks to previous research we know that ROR$\gamma$t  has no positive interactions with any of the transcription factors considered in the TRN and therefore lacks a feedback loop mediated by transcription factors [89]. The absence of a feedback loop regulating this master transcriptional factor could explain the dependence of Th17 cells on their environment. Thus, this missing link must be mediated by other signaling molecules that link the transcriptional response with the microenvironment.

These results show that the interactions among MTF are not sufficient to recover the configurations characteristic of CD4+ T cells types and highlight the importance of signaling pathways and the microenvironment. Given these results, we can continue to improve our model, creating a new version that takes into account the factors we suspect are missing.

**Fig. 3.11** Synchronous vs. asynchronous updating. Effect of the (**a**) synchronous or (**b**) asynchronous update in the final attractor. The nodes corresponding to the inputs of the system have value 0 and are not shown for clarity

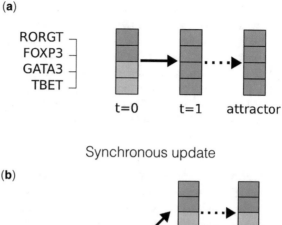

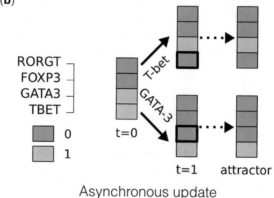

## Synchronous Versus Asynchronous Update of Boolean Networks

Before moving to a more realistic but also complicated model, let's use the simple model of interactions between MTF to illustrate the differences between synchronous versus asynchronous updates of Boolean models.

Let's suppose that there is a cell that expresses both T-bet and GATA-3 at the same time in the absence of external signals. Using the Boolean model we can try to determine the dynamical behavior of this cell (see Fig. 3.11a). We will consider that, as there are no external signals, the value of the inputs of the system is 0. The initial state $t = 0$ of this cell will be [1 1 0 0]. Evaluating all the Boolean functions, we can determine the fate of this cell. At the next time step $t = 1$ the state of the system will be [0 0 0 0] as both transcription factors will inhibit each other. If we evaluate the state [0 0 0 0] using the same functions, we can determine that at $t = 3$ the state of the system is [0 0 0 0] again. The state [0 0 0 0] is a steady state, and corresponds to the cell type Th0, where there are no MTF present.

However, an implicit assumption in this case is that we are evaluating all the nodes at the same time. This update method is called "synchronous." However, this assumption is not always true. For example, a node could be produced faster than the other affecting the other nodes. We can study this process by updating the nodes separately. This update method is called "asynchronous."

Using the same example, we can see the effect of the asynchronous update in the system (see Fig. 3.11b). Using the same initial state as in the previous example [1 1 0 0] we can determine the effect of the update schema. If we update TBET first, the node will be inhibited by the presence of GATA3 and reach the Th2 attractor [0 1 0 0]. Then, no matter which node we update, the system will stay in the [0 1 0 0] state; this means that [0 1 0 0] is a steady state of the system. On the other hand, if we update first the GATA3 node, it will be inhibited and the system will reach the Th1 attractor [1 0 0 0]. This contrasts with the synchronous update, as the same state can have two (or more) successors states and reach different attractors depending on the update order.

*Remark 3.12 (Updating Affected by Different Time-Scales)* Another case where the update time can be affected is when processes have different time-scales. For example, signaling can be faster than transcription [23]. In those cases, nodes can be updated according with their dynamic hierarchy [342].

## CD4+ T Cell Regulatory Network

As we have seen, including only the master transcription factors is not enough to explain the differentiation of CD4+ T cells. We can do better. To improve our model, we will now study a dynamical network that includes:

- Signaling pathways and their regulators.
- Cytokines that have been shown to be fundamental in CD4+ T cell type attainment.

Given the complexity of the new network and the high number of involved molecules in CD4+ T cell-fate attainment, we only show in Fig. 3.12 the IL-2 pathway. For the complete network see [303].

We can now proceed to add some new important information to the regulatory network shaped around the involved master transcription factors.

### Adding Signaling Pathways

When adding signaling pathways it is important to take into consideration post-transcriptional modifications and how they affect signal transduction. In a Boolean model we assume that a node is active (i.e., has the value of "1") when it can carry out its function. In the case of signaling pathways, this implies that not only the components of the network are expressed, but also that they have the necessary post-translational modifications (phosphorylation, complex formation,

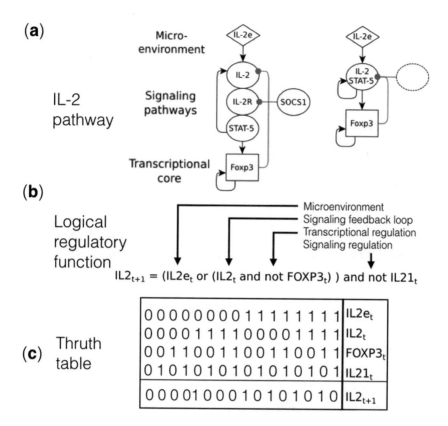

**Fig. 3.12** Networks can include more than one level of regulation as exemplified by the IL-2 pathway. (**a**) The network includes transcription factors (rectangles), signaling pathways (ellipses), and exogenous cytokines (diamonds). Activations between elements are represented with black arrows, and inhibitions with blunt arrows. These regulatory interactions can be simplified using mathematical methods (e.g., here we show that the linear pathway that is located at the center of the graph that represents the IL-2 pathway can be compacted to give rise to a new node). (**b**) The function of the IL2 node integrates multiple activators and inhibitors. (**c**) The truth table of the IL2 node was obtained from the function and represents available biological data

etc.) or spatial localization necessary to transduce the signal. For example, STAT proteins are usually expressed in CD4+ T cells, but they only transduce the signal to the nucleus if they are phosphorylated by a cytokine/receptor complex and dymerize with another STAT protein. In this case, for the STAT node to be considered to be active, a long chain of events is required. Including all the components and events associated can be computationally complex. To solve this problem there are network simplification methods that maintain the dynamics of the system while reducing the number of nodes (see for instance [340, 468]).

**Inclusion of Cytokines**

The cytokines in the cellular microenvironment are fundamental triggers for the differentiation of these adaptive immune cells. Further, as immune cells differentiate they also produce specific cytokines, altering the microenvironmental configuration. The same cytokine can be produced both by the cell it affects, but also by other cells in the tissue or cell culture. To distinguish between these two scenarios we separate the cytokines in two different nodes:

- Intrinsic cytokines, produced by the cells as they differentiate.
- Extrinsic cytokines, present in the microenvironment.

Since exogenous cytokines are part of the microenvironment, they are modeled as inputs of the system and cannot be regulated by the cell. Endogenous cytokines are intrinsic to the system and their production can be regulated by the differentiating cells, forming part of the feedback loops that determine the differentiation and maintenance.

**The Extended Model**

The resulting network includes:

- Transcription factors.
- Signaling pathways.
- Intrinsic and extrinsic cytokines.

Each signaling pathway is compressed into a single node that is active if the signal is transduced. The resulting network includes multiple levels of regulation:

1. The regulation in the nucleus by transcription factors.
2. The regulation by signal transduction pathways mediated by SOCS proteins.

*Remark 3.13 (Models can Return a Large Number of Attractors that Correspond to the Same Cell Type)* In the CD4+ T cell regulatory network represented in Fig. 3.12, there are some attractors that share the value of the intrinsic nodes but that differ in the value of the extrinsic nodes. Each of these attractors is a different solution of the system, but they correspond to the same cell type. At the same time, CD4+ T cells are highly heterogeneous. While each cell type has cellular markers and cytokines associated with it, there can be variations in which markers are expressed. Some of the attractors recovered by the model exhibit this behavior, where they express different expression profiles that can be biologically assigned to the same cell type. In this case, we used a criteria similar to biologist, where a cell is classified according to a master transcription factor and characteristic cytokines [521].

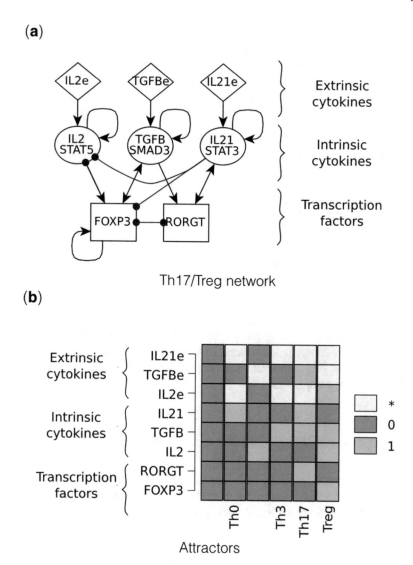

**Fig. 3.13** Th17/Treg network. Extrinsic cytokines, which are present in the environment, cannot be regulated by the cell. Intrinsic cytokines, produced by the cells as they differentiate, constraint transcriptional regulation. (**a**) shows the regulatory network, taking into account extrinsic cytokines as inputs. (**b**) Shows the attractors of the Th17/Treg network (i.e., Th0, Th3, Th17, and Treg)

## Multiplicity of Stable Configurations

The dynamical analysis of the network can yield a great number of stable configurations, as shown in Fig. 3.13b. However, most of these configurations are equivalent and can be classified into different subtypes to facilitate analysis. For example, the inputs of the network represent the cytokine microenvironment and are not used as markers to determine the cell phenotype. If we ignore the inputs, we can arrange the attractors according to the value of the molecular markers (transcription factors, intrinsic cytokines, etc.). Furthermore, some of these attractors are equivalent, as they can be classified as the same cell type.

To label the attractors of the network we can use biological criteria:

- Resting CD4+ T cells (labeled Th0) were defined as expressing no transcription factors or regulatory cytokines.
- Th17 was defined based on RORγt and STAT3 signaling mediated by IL-6 or IL-21, all of which require the presence of TGFβ.
- iTreg expressed Foxp3 and TGFβ, IL-10, or both, all of which require the presence of IL-2e.
- Tr1 was characterized by the presence of IL-10 .
- Th3 was characterized by the presence of TGFβ.
- Cells that express both cytokines (i.e., IL-10 and TGFβ) are labeled as IL-10+ TGFβ +.

The new model of the network recovers the attractors that correspond to: Th0, Th17, iTreg, and Th3 cells [521]. These results show that a network containing the transcription factors, signaling pathways, and intrinsic and extrinsic cytokines can recover some of the expression patterns observed in actual CD4+ T cells.

Once we know that the model recovers some of the biological behaviors we can begin to do more complicated tests.

## *Analysis of the CD4+ T Cell Regulatory Network*

In order to explore the validity of the resulting model, a procedure of analysis that tackles systems-level consequences of mutations and environmental disturbances, as well as phenotypic plasticity, is carried out.

### → **Mutants**

Boolean models allow us to simulate biological experiments in silico. These experiments can be used to verify the model by comparing its results with available experimental data. One of those experiments is simulating knockout or over-expression mutants. Knockout mutants are modeled by setting the value of the target node to 0, while over-expression mutants can be represented by setting the value of the target node to 1. One advantage of in silico models is that it is relatively easy to simulate the mutants and give predictions about experiments that have not been made in vivo.

In the case of our regulatory network Fig. 3.13 it is possible to obtain all the single-node mutants (Fig. 3.14a) and compare them with available information. Not all nodes are involved in all cell types, so mutating a node will only affect some of the possible attractors. We can then compare the patterns of loss and gain of cell types with the available experimental data. In this way, studying the mutants of the network lets us verify most of the interactions. Furthermore, not all the mutants have been studied in vivo. For example, some mutants like GATA3$^{KO}$ are lethal and require complicated conditional murine models. For some other cases, the effect of a mutation over a certain cell type has not been studied, for example, little is known about the recently described cell types like Th2 and Tr1. In this way, the mathematical model allows us to make predictions of various mutants where no experimental data are available.

To further verify the construction of the functions and the structural properties of the model, we can perform a robustness analysis altering the updating rules. For each topology there are many possible sets of functions. A possible problem in the construction of a network is over-fitting, where we chose a set of functions that retrieves the attractors we are searching for, but a small change in the functions can drastically search the resulting attractors invalidating the model. To verify the construction of the network we can alter the functions and determine how much these changes affect the results. As we have seen, we can express the function of a node as a rule or as a truth table. To study the robustness of the network, we alter some of its functions by randomly changing some of the values of their truth table (Fig. 3.14c). If the system recovers the same attractors it means it is stable to that perturbation, but if it loses or gains attractors, it means it is sensible to that perturbation. Given the number of possible perturbations of the truth table, it is necessary to do a random sampling of possible perturbations. This analysis allows us to test the robustness of structural properties of the networks to noise, mis-measurements and incorrect interpretation of the data. Another option is using model checking to determine all the possible sets of functions that recover the same attractors with the same topology (as discussed in [31]).

→ **Role of the Microenvironment**

When a CD4+ T cell is activated it will differentiate into different subsets depending on the cytokines in its microenvironment. Cytokines can be produced by the same cell (endogenous or intrinsic) or by other cells of the immune system or the organism (exogenous or extrinsic). In the model we included as inputs extrinsic cytokines. This allows us to study the relationship between cell types and exogenous cytokines.

We can study the effect of the microenvironment in CD4+ T cell-fate attainment by setting the values of the inputs according to the environments that have been defined experimentally (Fig. 3.14b). For example, if we want to simulate a regulatory pro-Treg environment, we can set the value of IL2e and TGFBe to 1, and set the value of the other inputs as 0. On the other hand, if we want to study an inflammatory environment like pro-Th17, we can set the value of IL21e and TGFBe to 1, and set the value of the other inputs as 0. One advantage of this methodology is that it also helps us simplify the computational problem. For $N$

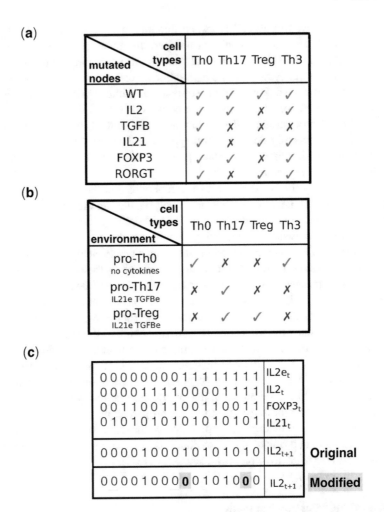

**Fig. 3.14** In silico experiments for the Th17/Treg network. (**a**) Knockout in silico experiments for the Th17/Treg network. We simulated loss of function or null mutations (KO) by setting the function of the target node to 0 and determine the resulting attractors. Mutating some nodes like TGFB can cause the loss of multiple cell types, while some cell types are very robust to mutations like Th0. The ticks represent attractors that were recovered and crosses attractors that were lost. (**b**) Attractors obtained in the different microenvironments. (**c**) Truth table of the IL2 node without and with random perturbations

inputs there are $2^N$ possible combinations of inputs, however, most of them are not biologically relevant. By focusing on the biological relevant microenvironments we obtain relevant information that can be compared with experimental data and focus our research in relevant biological situations.

Here, we focus on only the most relevant microenvironments: pro-Th0, pro-Th17, and pro-Treg (Fig. 3.14b). Then, we determine which cell types can be recovered in each environment, and show that Treg cells require IL2e, and Th17 cells require TGFBe, as can be seen by the resulting attractors (Fig. 3.14b). Th0 and Th3 can be maintained in the absence of extrinsic cytokines as can be seen in the pro-Th0 environment. In a pro-Th17 environment we can only recover Th17 cells, but in a pro-Treg environment we can recover both Treg and Th17 cells (Fig. 3.14b). The recovered behaviors agree with the experimental data and also with previous models [2, 521]. Furthermore, the coexistence of Treg and Th17 cells in a pro-Treg environment is associated with chronic association, which we will discuss later.

Until now we have supposed that the signal in the microenvironment is constant. However, this is not necessarily true. The signals in the environment, signal transduction, and expression of transcription factors are subject to temporal changes. For example, an exogenous cytokine could be produced only for a certain period by cells of the immune system, creating a peak in its expression. We can study this kind of phenomena by transiently changing the value of the nodes (Fig. 3.15). For example, in this model a temporal activation of the TGFBe node is enough to change the cell from a Th0 to a Th3 phenotype. However, transient perturbations of the IL2e or the IL21e nodes are not enough to transition towards Th17 or Treg, as we require constant signaling of these nodes to maintain the phenotype.

## → Plasticity

CD4+ T cells also exhibit phenotypic plasticity and memory. This means that, once differentiated, their new expression pattern can often be maintained even after the removal of the (microenvironmental) triggers of differentiation. This plastic response has been associated to the capacity to robustly adapt to changes in the immune challenges. There are various ways to study this plasticity. In this section we will focus on the transitions between attractors caused by transient perturbations in the values of the nodes.

The cytokines in the microenvironment, the activation of signaling pathways, and the expression of transcription factors are not always constant; they are subjected to noise and small perturbations. We can determine the effect of these perturbations in the cell-fate attainment by transiently perturbing the value of the attractors of the system. These transient perturbations in the values of the nodes are equivalent to developmental noise, temporal changes in the microenvironment, risk factors, or clinical interventions. For example, if we have a regulatory Treg attractor, we can transiently activate the value of the IL21e node for a time step, and determine the effect of the perturbation (see Fig. 3.15b). In this case the system transitions towards a Th17 attractor, showing that transient expression of IL-21 in the microenvironment can shift the system towards an inflammatory response, even in a pro-Treg environment. Whether this response is detrimental or not depends on the circumstances. In case of an infection a robust immune response is necessary to control the pathogen, but if the patient is healthy it can lead to a chronic inflammation.

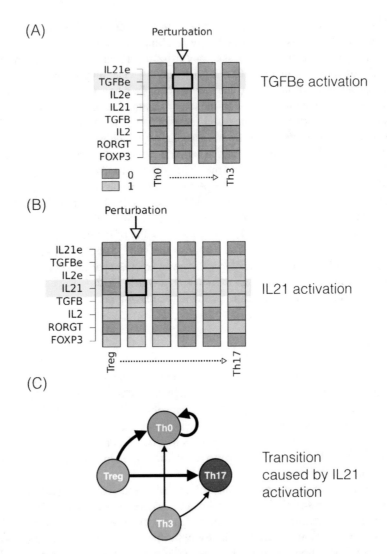

**Fig. 3.15** Plastic transitions in response to transient perturbation in the value of the nodes. (**a**) The transient perturbation of the TGFBe node causes a transition from a Th0 to a Th3 attractor. As TGFBe is an input we set its value to 1 during one time step and then return the node to its original value of 0. (**b**) The transient perturbation of the IL21 node causes a transition from a Treg to a Th17 attractor. We set the value of IL21 value to 1 during one time step and then return the node to its original function. (**c**) Transitions caused between cell types in response to the transient one step of IL21

We can do this experiment for the same node for each attractor to study the secondary effects of the perturbation (see Fig. 3.15c). If the cell returns to the same attractor we say it is stable to the perturbation, but if it transitions to a new attractor we say it is plastic to that perturbation. Some of the perturbations do not cause transitions, but let the system return to the original cell type. In this way, we can achieve a measure of the stability of a given cell type. This shows that the regulatory network generates restrictions in terms of cell types but also in terms of the patterns of cell-fate transitions. If this analysis is repeated for every node of every attractor, the result is a cell-fate map where the nodes represent CD4+ T cell types recovered by the network and the connections represent the possible transitions between pairs of differentiated cell types. Some of these transitions are more common than others, and other transitions are only possible in certain microenvironments.

*Remark 3.14 (The Plasticity of CD4+ T Cells and Its Microenvironment)* The microenvironment also affects the plasticity of CD4+ T cells, as the inputs of the system limit the possible transitions between cell types (Fig. 3.15b). In general, if a microenvironment favors a certain cell type, the cell type will be more stable (as the system will return to it after a transient response to perturbations) and there will also be more transitions towards that cell type. However, other attractors are still reachable, and there exist transitions from and towards them. These "cell fate maps" highlight the complexity of the immune response, where there is a high diversity of cell types coexisting together.

Until now we have focused on Boolean models to study CD4+ T cells. However, the differentiation and plasticity of these cells is also affected by the concentration of the cytokines in the environment. As we have seen in previous chapters there exist multiple approaches that use ordinary differential equations to study biomedical systems. As we are focused on how the structure of the network affects the phenotype of these cells, we will use the approach used in [19, 112] to convert Boolean models to continuous functions.

## Continuous Model

In Boolean models both the value of the nodes and the time steps are discrete. However, in a ordinary differential equations approach, they are continuous. This means we can obtain intermediate concentrations of transcription factors and cytokines through time. The approach we are using recovers both regulatory (Fig. 3.16a) and inflammatory (Fig. 3.16b) cell types, where the final value of the nodes tends to 0 or 1. However, we can also observe transient peaks of expression and cells that achieve intermediate phenotypes, which have been experimentally observed [127, 521] (Fig. 3.16c).

*Remark 3.15 (Assessing How Different Concentrations of Cytokines Affect Cell-Fate Attainment)* Continuous approaches can be used to determine how different concentrations of two (or more cytokines) affect cell-fate attainment. We can estimate the concentration of cytokines necessary for a cell to differentiate into a regulatory or inflammatory cell type and how the presence of other signals in the environment will affect this. For example, while high concentrations of regulatory

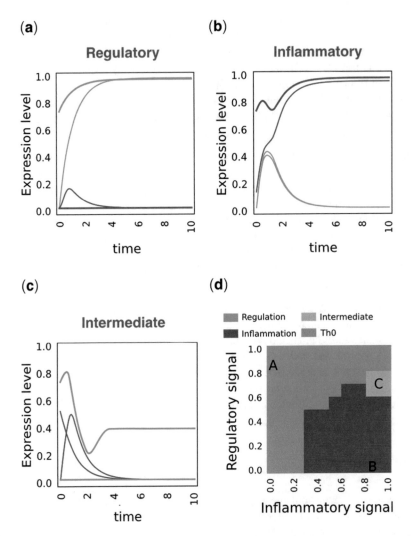

**Fig. 3.16** Continuous approximations allow to study the effect of the concentration or expression level of different nodes in cell fate attainment. Depending on the value of the regulatory and inflammatory exogenous cytokines in the environment, the model can converge to a (**a**) regulatory, (**b**) inflammatory, or (**c**) intermediate cell fate. (**d**) The relationship between different exogenous cytokines—that serve as inputs to the system—can be visualized in a bifurcation diagram

cytokines will induce regulatory cell types and high concentrations of inflammatory signals will induce inflammatory cell types, high concentrations of both regulatory and inflammatory cytokines can induce cells with intermediate phenotypes that have been associated with chronic inflammation (Fig. 3.16c, d).

## *Integrating Other Systems*

As we have discussed, the immune system strongly affects, and is affected by the rest of the organism. In particular, there is a close communication between the metabolism and the immune system, which underlies the feedback loop between obesity and chronic inflammation. However, a limitation for creating Boolean models that join metabolism and inflammation is that we lack detailed molecular information of how both systems communicate. Now, thanks to the constant advances by experimental biologists, we begin to understand the pathways that mediate the information exchange between both systems.

To finish this section, and in order to illustrate the versatility of the proposed model, we will show in what follows a small example of how hyperinsulinemia affects CD4+ T cell-fate attainment, favoring inflammatory responses.

### → **Hyperinsulinemia**

Hyperinsulinemia is characterized by an increase in the levels of insulin and is associated with metabolic syndrome. High levels of insulin inhibit the regulatory cytokine IL-10 through the Akt/mTOR pathway [191]. Using this information, we can expand the CD4+ T cell model. First, we determine the signaling pathways, then we find the nodes in common and integrate both networks.

In this case, hyperinsulinemia acts as an input of the network, as it is an exogenous factor that regulates the network. As we have seen, CD4+ T cells are plastic and dynamically change from one type to others, depending on the microenvironment and transient perturbations or initial conditions. To explore this, we can obtain both the attractors and the cell-fate map in different microenvironments with or without hyperinsulinemia.

The model provides an explanation to some paradoxical behaviors observed in CD4+ T regulatory cell populations during obesity-associated chronic inflammation. TGFβ can promote both inflammatory Th17 cells and regulatory Tregs, and transitions between both subsets have been observed. TGFβ is necessary for the differentiation of both subsets, and transient signaling via the STAT3 pathway may be enough to shift some cells towards Th17, as the model shows. In obesity, Tregs expression profiles are similar to inflammatory T cells and transfer and depletion of adipose Treg cells have been reported to both improve or worsen insulin sensitivity, depending on the model and the population studied [283, 310, 354]. Such apparently paradoxical behaviors can be explained by the relationship between TGFβ and IL-10 in the context of the dynamic regulatory network model used here. Under hyperinsulinemia, Th17 cells become more stable while IL10+ cells are lost. The remaining regulatory cells express TGFβ that is involved in Th17 differentiation, while insulin alters iTregs stability. In this way, the model predicts that hyperinsulinemic inflammatory environments' regulatory T cells are less stable.

*Remark 3.16 (Exploring Therapeutic Interventions)* The model also allows us to propose therapeutic interventions. For example, if we have a patient with chronic inflammation we may be interested in which signals can cause a transition from an inflammatory to a regulatory effect. The model shows that the transient activation of IL-10 can make inflammatory cells like Th17 transition towards regulatory IL10+TGFB+ cells as can be seen in Fig. 3.17. Furthermore, the model can predict secondary effects of this intervention by simulating the same perturbation in other cell types. We focus on transitory perturbations as we want to return the system towards a healthy system where it can react to the signals of the environment, including inflammatory signals caused by pathogens. If there is a permanent increase of the concentration of IL-10 in the microenvironment this may cause immunosuppression in the patient. The model also allows us to determine risk factors and their effects. Peaks in the concentration of insulin as those observed in hyperinsulinemia cause transitions towards inflammatory cell types. In this way, metabolic dysregulation can affect the equilibrium of the immune system.

As can be seen, the proposed modeling approach opens the door to the construction of mechanistic explanations of complex disease circumstances.

**Fig. 3.17** Effect of hyperinsulinemia in CD4+ T cell fate plasticity. Transient increases in the level of IL-10 or insulin can cause phenotypic transitions towards regulatory or inflammatory cell types, respectively. The width of the arrow represents the frequency of the transition

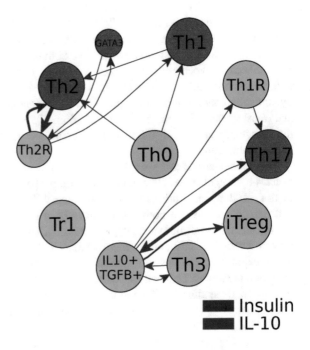

## *Final Comments and Perspectives*

For healthy conditions to be maintained, the immune system has to *dynamically respond and cooperate* with the microbiome and communicate with other parts of the organism that create complex environments. These environments are constantly changing. Also, the immune system is subjected to developmental noise. To adapt, the immune system has to be (depending on the circumstances) both:

- Plastic.
- Stable.

Grounded on experimental data, mathematical models help us understand how the molecular mechanisms that underlie cell-fate attainment achieve this complex dynamical behavior.

Taking chronic inflammation as a modeling subject, in this section we have seen how Boolean regulatory networks serve us to:

1. Review the available information.
2. Determine which areas need more research.
3. Find common patterns in the regulatory network.

Discrete Boolean networks also let us validate the network against biological data and construction errors. Once the network has been constructed and validated, it is possible to study its dynamics to recover:

- The differentiation patterns.
- The plasticity characteristics.
- The effect of the microenvironment.

Furthermore, models let us look at systemic behaviors, not focusing on only one molecule, cell type, or transition, but integrating all possible behaviors in a formal and unified system.

This bottom-up integrative approach allows us not only to understand the system we are studying, but also its relationship with the rest of the organism. For example, we can integrate the effect of hyperinsulinemia in CD4+ T cells, explaining complex patters observed in obesity-associated chronic inflammation. In this way, system biology gives us a useful toolbox to unravel the complex relationships between metabolism and the immune system.

We hope that the methodology presented here is useful for both experimental and theoretical scientist to simulate, validate, and analyze complex biological systems. In what follows, we shall tackle the study of complex phenomena that will require a quantitative modeling approach: *Atopic dermatitis*.

## 3.4   Atopic Dermatitis

Atopic dermatitis is a complex disease. First, because several combinations of genetic and environmental risk factors can trigger the onset of this disease, and second, because there are *many different stages of the disease*, with increasing severity. Efficient preventive treatments should thus aim to halt the progression of the disease from a mild and asymptomatic phenotype to severe forms that are difficult and costly to treat.

Further, such intervention strategies should ideally be effective for the whole spectrum of disease phenotypes. Until recently, understanding the mechanisms underlying the onset, progression, and prevention of this disease had been difficult because of the complex interconnections that exists between the many risk factors and the associated disease phenotypes. In this section, we show some recent mathematical models of atopic dermatitis that have shed light on the mechanisms for onset, progression, and prevention of this disease, providing plausible answers to clinically relevant questions, such as the design of optimal and personalized treatment regimens.

We follow here a quantitative modeling approach, and the proposed mathematical models consists of systems of ODEs (introduced in Sect. 2.6), some of which operate at different time-scales to describe the complex interplay that exists between fast biochemical reactions and slower tissue-level processes, as described in Sect. 2.9.

Let's proceed with our example.

### *Motivation*

Atopic dermatitis is a skin disease characterized by a defective epidermal permeability barrier function that appears as dry and scaly skin, and by aberrant immune responses to environmental insults, manifested as excessive inflammation and allergy [156]. This disease affects approximately:

- 15% of infants worldwide;
- 20% in the United Kingdom [474];
- 15% in Germany [128];
- 10% in Nigeria [349].

And its incidence has been rapidly increasing (see for instance [128, 128, 474]). Moreover, patients with a clinical history of atopic dermatitis have a strong predisposition for developing other atopic diseases, such as asthma and allergic rhinitis [95, 156, 517].

In terms of its economic impact, the average costs of treating atopic dermatitis per patient per year has been estimated to represent up to 4480 USD in the United States [136] and 1425 EUR per patient per year in Germany [128]. These costs are increasing with the augmenting prevalence of atopic dermatitis [297].

Despite its clear socioeconomic relevance, the mechanisms leading to atopic dermatitis have not been fully elucidated, limiting the treatment options to relieve symptoms [156, 387, 436]. For instance:

- emollients enhance the permeability barrier function [99];
- steroids decrease the inflammation [351];
- antibiotics reduce the infection that results from a defective epithelial function [270].

However, long-term treatment does not guarantee remission of the disease [441], and can even lead to aggravation of the atopic dermatitis condition by further affecting the epidermal structure [107, 351, 407].

Finding effective treatment options for atopic dermatitis has been challenging because of following four features of this disease:

1. There are *several* different predisposing genetic [35, 143, 208, 250, 356, 389, 444] and environmental [35, 95, 99, 134, 236, 492] risk factors that have been linked to the development of atopic dermatitis.
2. The risk for developing atopic dermatitis as a consequence of predisposing factors is often dose-dependent. Only a severe genetic deficiency [66], or high amounts of environmental triggers [338, 403] can trigger the onset of the disease.
3. There is synergism between risk factors, because the presence of two or more risk factors dramatically increases the susceptibility to develop atopic dermatitis, in a non-additive way [337, 403].
4. The pathogenesis of atopic dermatitis comprises several different phases, characterized by distinctive epidermal phenotypes of increasing severities [270, 436].

Together, these observations suggest that effective treatment strategies must account for the risk-factor and stage-dependent, pathogenic process of the disease, which might be specific for different patient cohorts.

These clinical challenges all arise from the fact that these predisposing risk factors are strongly interlinked, forming a complex network of cellular and biochemical interactions prone to perturbations by multiple risk factors (recall Fig. 1.1). Different risk factor combinations act, interact, and propagate across the network in unanticipated ways, resulting in:

- The existence of multiple possible combinations and strengths of perturbations that can converge to a limited number of disease phenotypes of atopic dermatitis (fragmentation of the phenotypic space, illustrated in Fig. 2.22).

- The propagation of disturbances across the regulatory network, which are responsible for the gradual aggravation of the atopic dermatitis phenotype (Fig. 1.4).
- The non-additive interactions between risk factors due to the intricate topology of the feedback control structure underlying phenotypic determination.
- The nonlinear effects of risk factor severities on the phenotypic transitions (Fig. 2.21).

Therefore, mapping risk factor combinations to the healthy and pathological disease phenotypes requires the systems-level, regulatory network approach that has been discussed throughout this volume. Further, understanding the mechanisms of disease progression, from a mild asymptomatic to a severe and treatment-resistant phenotype, is fundamental to devise preventive treatment strategies, and requires an explicitly dynamical systems approach. Indeed, one of the fundamental motivations of proposing and analyzing dynamical models of disease is to prevent or decrease the incidence of late-stage diseases. The reason for this is that treating late stages of chronic diseases requires a higher treatment effort with an associated increased cost, and with increased negative side effects. For example, while the early and asymptomatic stages of atopic dermatitis can be treated with the application of emollient creams that have a low economic cost and no known negative side effects [436], advanced stages of atopic dermatitis require continuous applications of corticosteroids, which are costly and have associated negative side effects (tissue atrophy) [387, 390, 418]. Consequently, it is also important to find the minimal treatment strengths that lead to the remission of advanced disease stages. Systems biology approaches can be used for such an optimization of treatment regimens, as has been shown for:

- prostate cancer [198–200, 407];
- pneumococcal infection [121];
- and indeed also for atopic dermatitis [88, 438].

Recently, we and our collaborators proposed a series of mathematical models of atopic dermatitis that give answer to these clinical problems from a systems biology perspective. These models were constructed and analyzed using some of the methods and techniques presented throughout this volume (Sects. 2.6 and 2.9), and thus illustrate how different modeling methods can be applied to answer specific clinical questions.

This section is organized as follows:

**Primo:**     We give a brief overview of how the different genetic and environmental risk factors are connected among them by a complex reaction network that under healthy conditions controls epidermal homeostasis and that is disrupted in atopic dermatitis.

**Secondo:**     We introduce three mathematical models with increasing complexity, which cover growing regions of these reaction networks, and explain the insights gained from these models:

- First, we will describe two models representing the innate and the adaptive immune responses that govern the onset and progression of atopic dermatitis, which are abruptly activated by the presence of environmental stressors and other immune components, respectively.
- Then, we will couple the dynamical model of innate immune responses to slower tissue-level dynamics, and show how different risk factors mediate the turning-on-and-off dynamics of these innate immune responses.

**Trezo:**   We explain the clinically relevant questions addressed with these models.

- How tissue-damaging, chronic inflammation is established as a consequence of frequent or long-lasting activations of innate immune responses, and underlie the progression of atopic dermatitis.
- How the last model, which describes the progression from a mild and asymptomatic to a severe atopic dermatitis phenotype, has helped to understand how the worsening of this disease can be prevented.
- Finally, we will briefly discuss how the proposed model can be used as a quantitative framework to find optimal treatment options to revert severe symptoms of atopic dermatitis with the minimal amount and duration of corticosteroids and emollients. For this, we will refer to recent publications that have tackled this issue, specifically: [88, 438].


## The Biology of Atopic Dermatitis: Regulatory Network Controlling the Complex Interplay Between Hallmarks and Risk Factors

> **Note**: This subsection was taken from [120]:
>   URL: http://hdl.handle.net/10044/1/47969
>   (published under a Creative Commons Attribution
>   Non-Commercial No Derivatives License
>   https://creativecommons.org/licenses/by-nc-nd/3.0/).

The phenotype of atopic dermatitis is characterized by three hallmarks:

1. A dysfunctional skin barrier [133].
2. Propensity for infection by bacteria such as *Staphylococcus aureus* [251, 309]. and:
3. Frequent [270] and long-lasting [255] inflammation that is sometimes accompanied by allergic reactions to ubiquitous environmental insults [176].

Many genetic and environmental risk factors have been associated with an increased propensity to develop atopic dermatitis, all of which affect these three hallmarks of this disease, either directly or indirectly.

Skin barrier function is affected by genetic or environmental risk factors that decrease the expression of structural components of the skin barrier, most notably the protein *filaggrin* [143, 306, 356] that is responsible for the cross linking between the extracellular lipid envelope and the intercellular cytoskeleton of the keratinocytes that form the skin barrier [397], and has been identified as a key determinant of the permeability barrier function of the epidermis [143]. Genetic risk factors that decrease the expression of filaggrin and other components of the skin barrier are polymorphisms [66, 361] and mutations [143]. Environmental factors can also impair the expression of barrier function components, for example, by the prolonged use of hard water [98], which interferes with the calcium gradient that regulates the expression of terminal differentiation markers of keratinocytes [130, 131, 400, 456]. Also chronic inflammation, which is actually a characteristic feature of advanced atopic dermatitis, is associated to a decrease in the expression of barrier function components [52, 117, 130, 209, 350]. Immune responses are affected by risk factors that alter the concentrations or activities of different components of their mediating signaling cascades [12, 35, 309, 345, 444]. These changes can result from polymorphisms [309, 444] or environmental fluctuations, such as changes in the *microbiota* [35]. An important mediator of both immune responses [65, 68, 423] and barrier function [76, 184, 423] are the networks of kallikrein proteases, the activity of which can be altered by further environmental and genetic risk factors, such as the frequent use of soaps and detergents [98] that raise the pH, increasing the catalytic activity of these enzymes [60], the genetically determined decrease in the expression of the protease inhibitor LEKTI [208, 389], or the increased expression and activity of the protease [33, 250].

Not only the risk factors, but also the resulting symptoms of atopic dermatitis are associated to the hallmarks of this disease: dysfunctional skin barrier leads to an increased permeability to environmental factors, such as pathogens, thereby increasing the susceptibility for infection [113, 133]. In turn, increased pathogen load in the viable epidermis weakens the skin barrier, since pathogens interfere with the barrier repair mechanisms [185, 266]. Moreover, augmented pathogen loads in the viable epidermis triggers further inflammatory flares of atopic dermatitis [113]. Excessive activation of innate immune responses can lead to the activation of Th2-mediated adaptive immune responses [65, 108, 257], which contribute to the establishment of a pro-inflammatory microenvironment that further impairs barrier function by interfering with gene expression programs that control the barrier remodeling process [350, 448], and underlies the allergic reactions that characterize severe forms of atopic dermatitis [160, 168, 347].

This internal regulatory logic of interconnected reaction network elements (Fig. 3.18) is responsible for:

1. Phenotypic convergence.
2. Disease aggravation.
3. Synergism between risk factors.

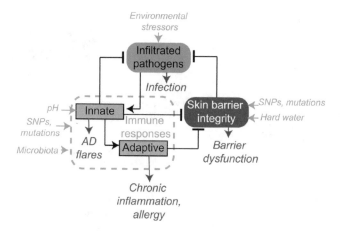

**Fig. 3.18** Interplay between hallmarks (red) and risk factors (blue) of atopic dermatitis, interconnected by the regulatory interplay between pathogens, innate and adaptive immune responses, and skin barrier

In what follows we will present increasingly complex mathematical models that represent this regulatory interplay between:

- pathogens,
- innate and adaptive immune responses,
- and skin barrier.

## Modeling Immune Responses as Bistable Switches

One of the hallmarks of atopic dermatitis is the aberrant immune responses, which are responsible for the typical flares of atopic dermatitis (innate immune responses) and of the establishment of chronic inflammation and allergic reactions (adaptive immune responses). These were mathematically represented and analyzed with the ordinary differential equations models of Tanaka et al. [439] and Hoefer et al. [201], which will be explained in the following lines.

### → Modeling Innate Immune Responses with a Reversible Switch

In atopic dermatitis, a key player mediating the innate immune responses to pathogens and other environmental stressors that have infiltrated into the viable layers of the epidermis (Fig. 3.19a) is the Protease-Activated receptor (PAR2) pathway. The activity of this signaling cascade is mediated by a complex interplay between proteases called kallikreins (KLK), PAR2, and the kallikrein inhibitor LEKTI

**Fig. 3.19** Modeling innate immune responses in Atopic Dermatitis. (a) Innate immune responses in atopic dermatitis (b) are mediated by a complex network of Protein-Protein-Interactions controlling the activity of PAR2 in response to pathogen challenges. (c) The abrupt onset and cease of flares of atopic dermatitis in response to pathogen stressors is described by a bistable dose–response behavior, which is affected by genetic and environmental risk factors for atopic dermatitis. Figure adapted from [439] with permission from the Author; DOI of the original manuscript: https://doi.org/10.1371/journal.pone.0019895 (published under the Creative Commons License https://creativecommons.org/licenses/by/4.0/)

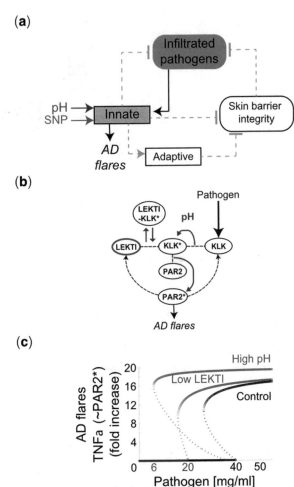

(Fig. 3.19b). Two major risk factors for atopic dermatitis affect the functioning of this reaction network:

1. Changes in pH, which affect the catalytic activity of the KLKs and the affinity of the LEKTI to its inhibitory target KLK.
   and:
2. Genetically determined decreases in the expression levels of LEKTI.

To explore the impact of these risk factors on the PAR2-mediated immune responses to pathogens, Tanaka et al. proposed in 2011 an ordinary differential equations-based mechanistic and kinetic mathematical model of this complex Protein-Protein-Interaction network [439].

The mathematical model is a set of six ordinary differential equations representing the experimentally described dynamic interplays between inactive (KLK) and active (KLK$^*$) protease, inactive (PAR2) and active (PAR2$^*$) protease receptor, unbound protease inhibitor (LEKTI), and the inhibitory complex (KLK$^*$LEKTI):

$$\frac{d[KLK^*LEKTI]}{dt} = k_a[KLK^*][LEKTI] - k_d[KLK^*LEKTI] - \delta_{LK}[KLK^*LEKTI],$$

$$\frac{d[LEKTI]}{dt} = -k_a[KLK^*][LEKTI] + k_d[KLK^*LEKTI] + t_L(m_L + f_L[PAR2^*])$$
$$-\delta_L[LEKTI],$$

$$\frac{d[KLK^*]}{dt} = -k_a[KLK^*][LEKTI] + k_d[KLK^*LEKTI] + k\frac{[KLK^*][KLK]}{[KLK^*]+C_K}$$
$$-\delta_{K^*}[KLK^*],$$

$$\frac{d[KLK]}{dt} = -k\frac{[KLK^*][KLK]}{[KLK^*]+C_K} - \delta_K[KLK] + f_{KS}S + f_K[PAR2^*],$$

$$\frac{d[PAR2]}{dt} = -k_P\frac{[KLK^*][PAR2]}{[KLK^*]+C_P} - \delta_P[PAR2] + m_P,$$

$$\frac{d[PAR2^*]}{dt} = k_P\frac{[KLK^*][PAR2]}{[KLK^*]+C_P} - \delta_{P^*}[PAR2^*].$$

(3.1)

Using the Law of Mass Action (Sect. 2.6), this system of ODEs formally represents the following reactions:

- The reversible formation ($k_a[KLK^*][LEKTI]$) (i.e., with a non-zero dissociation term $k_d[KLK^*LEKTI]$) of the inhibitory complex ([KLK$^*$LEKTI]).
- The degradation rates of all the molecules involved ($d_x[X]$).
- The proteolytic activation of KLK ($k\frac{[KLK^*][KLK]}{[KLK^*]+C_K}$), and of PAR2 ($k_P\frac{[KLK^*][PAR2]}{[KLK^*]+C_P}$). and:
- The de novo expression of LEKTI ($t_L(m_L + f_L[PAR2^*])$), KLK ($f_{KS}S + f_K[PAR2^*]$) and PAR2 ($+m_P$).

Two feedback mechanisms regulate the activity of the network:

- The auto-catalysis of KLK (first positive feedback).
     and:
- The active PAR2-mediated production of KLK5.

As far as the interaction with the environment is concerned:

**Input to the system:**   Given by the infiltrated pathogens ($S$) that induce the production of inactive KLK5.
**Output of the network:**   Given by the active PAR2 ([PAR2$^*$]), which drives the release of antimicrobial peptides and pro-inflammatory cytokines that characterize the flares of atopic dermatitis.

The mathematical model also considers the environmental and genetic risk factors known to affect the functioning of this network:

- High pH, which dramatically increases the catalytic activity of active KLK [60] and its affinity for LEKTI [118]. It is represented in the model by an increase in the catalytic rates of KLK* (parameters $k$ and $k_P$ in Eq. (3.1)) and the affinity between active KLK and LEKTI (parameters $k_a$ and $k_d$ in Eq. (3.1)).
- Decreased expression of the KLK inhibitor LEKTI [76, 152, 389], is represented in the model by a lower LEKTI production rate (parameter $t_L$ in Eq. (3.1)).

Analysis of the model was centered around understanding:

- how the flares of atopic dermatitis result from infiltrated pathogen loads,
- how this input–output relation is affected by the genetic and environmental risk factors.

For this, the dose–response behavior between pathogen load (input) and the *stable steady-state values* (see Box 2.2 in the previous chapter) of PAR2 activity (output) was assessed, neglecting the fast transient dynamic behavior of the Protein-Protein-Interactions.

The resulting *bifurcation diagram* (recall its schematic representation in Fig. 2.21) displays a robust bistable behavior, with on-and-off-states of the switch corresponding to the abrupt onset and cease of the flares of atopic dermatitis.

Genetic and environmental risk factors dramatically increases the propensity to develop the flares in response to pathogenic challenges by decreasing the threshold for onset and increasing the threshold for ceasing of atopic dermatitis flares, respectively (Fig. 3.19c).

### → Modeling Adaptive Immune Responses with an Irreversible Switch

Advanced and more severe forms of atopic dermatitis are characterized by chronic inflammation and allergy [436]. A key event in the establishment of these pathogenic features is the infiltration of Th2 cells into the epidermis [517]. This process is preceded by the polarization of naïve CD4+ T cells into Th2 cells, and corresponds to an irreversible differentiation event that is controlled by the master transcriptional regulator of CD4+ T cells polarization, Gata3. This regulator is induced by pro-inflammatory cytokines, such as IL4, which are released by dendritic cells that have migrated from the epidermis to the lymph nodes in response to previous innate immune responses [160]. A first step to understand the progression of atopic dermatitis from a mild phenotype to the establishment of severe symptoms in the form of allergic inflammation and allergy is thus to elucidate how CD4+ T cells polarization occurs in response to IL4 (Fig. 3.20a).

In [201], Höfer et al. proposed the first mathematical model of irreversible Th2 polarization in response to the increased expression of Gata3 by IL4. This simple mathematical model describes the dynamics of the Gata3 activity in the

**Fig. 3.20** Adaptive immune responses in Atopic dermatitis. (**a**) Adaptive immune responses in atopic dermatitis (**b**) are mediated by the Gata-3-dependent activation polarization of naive T cells in response to stimulation with the pro-inflammatory cytokine IL4. (**c**) This irreversible T cell polarization is described by a bistable switching behavior, which emerges from the underlying reaction network controlling Gata-3 activity that displays positive feedback and cooperativity. Figures (**b**) and (**c**) taken from [120] (URL: http://hdl. handle.net/10044/1/47969, published under a Creative Commons Attribution Non-Commercial No Derivatives License https:// creativecommons.org/ licenses/by-nc-nd/3.0/)

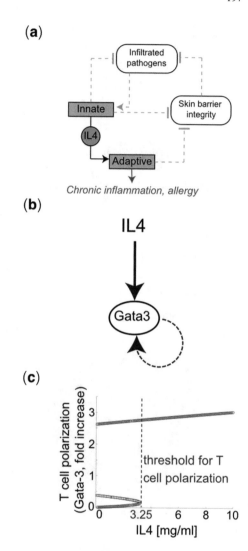

undifferentiated T cells that reside in the lymph nodes, in response to stimulation with IL4, as:

$$\frac{d[\text{Gata3}(t)]}{dt} = \alpha[\text{IL4}] + \frac{\kappa_G[\text{Gata3}(t)]^2}{1 + [\text{Gata3}(t)]^2} + \kappa[\text{Gata3}(t)].  \tag{3.2}$$

In this model, Gata3 expression is mediated by:

- An IL4-dependent de novo production term ($\alpha[\text{IL4}]$).
- A positive feedback term describing the Gata-3 mediated Gata-3 expression ($\kappa[\text{Gata3}(t)]$).
- A nonlinear term ($\kappa_G[\text{Gata3}(t)]^2/(1 + [\text{Gata3}(t)]^2)$) that represents the post-translational modifications of Gata-3 (Fig. 3.20b).

Steady-state simulations of this one-dimensional ordinary differential equation
(Eq. (3.2)) show an irreversible, bistable dose–response behavior (with a cease
threshold value < 0, Fig. 2.21) that characterizes the irreversible polarization of
individual Th2 cells (Fig. 3.2c).

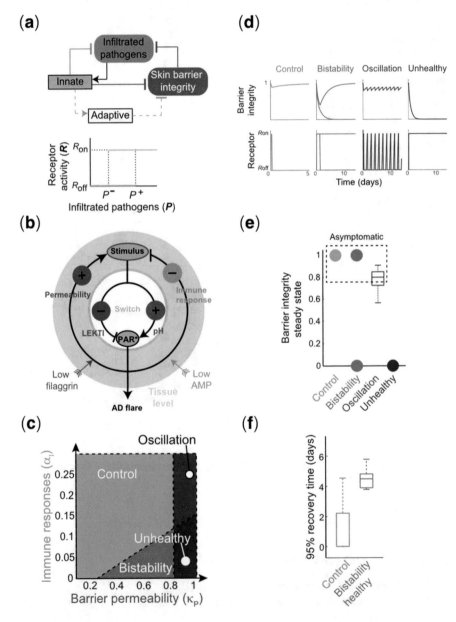

**Fig. 3.21** (continued)

## The Flares of Atopic Dermatitis Result from the Interplay Between Biochemical and Cellular Networks

In a previous section, we introduced the model of Tanaka et al., which reproduces the switch-like dose–response relation between infiltrated pathogens and the PAR2-mediated flares of atopic dermatitis (Fig. 3.19c). There, the input to this system, i.e., the infiltrated pathogen load ($S$), was treated as bifurcation parameter that can be swept in an autonomous manner. However, the infiltrated pathogen not only controls the onset and cease of the flares, but is also modulated by other tissue-level consequences of the activation of the PAR2-mediated signaling pathways (Fig. 3.21a). On the one hand, the activation of innate immune responses result in the weakening of the skin barrier through the activation of skin-barrier-degrading KLKs [184] and inhibition of barrier-restoration processes, such as the release of lamellar bodies into the corneal layer of the skin [117]. A defective skin barrier with low barrier integrity allows more exogenous stimuli to invade the inner epidermal layers, forming a positive feedback loop from the onset of atopic dermatitis flares (PAR2*) to the stimulus concentration. On the other hand, the innate immune responses triggered by PAR2* also contribute to the eradication of the accumulated stimulus in the inner epidermal layers by mediating the release of antimicrobial peptides or the induction of keratinocyte phagocytosis [1, 124, 266, 414], forming also a negative feedback loop from PAR2* to the stimulus concentration. The concentration of stimulus that penetrates the inner epidermal layers is thus determined by the balance between the positive and negative feedback regulations, whose strengths respectively depend on the skin permeability and *the capacity of stimulus eradication*, both of which are tissue-level processes that occur at a slower time-scale than the biochemically determined immune response switch (Fig. 3.21b).

Two of the mayor genetic risk factors for atopic dermatitis directly affect the strength of these feedbacks. A decreased filaggrin expression, caused by mutations or polymorphisms [66, 222, 245], weakens the skin barrier, increasing its permeability to environmental stimuli [242]. In turn, a decreased expression of immune system components [364] results in lower levels of pathogen-eradicating Anti-Microbial

---

**Fig. 3.21** Modeling the early phases of Atopic Dermatitis. (**a**) The early phases of atopic dermatitis are characterized by the interplay between infiltrated pathogen load, skin barrier integrity and reversible-switch-like innate immune responses. (**b**) The regulatory network underlying early phases of atopic dermatitis is a multi-scale structure in which the flares are controlled by the interplay between fast-switch-like biochemical processes and slow tissue-level dynamics. Figure adapted from [123] DOI: http://dx.doi.org/10.1098/rsfs.2012.0090 (published under the Creative Commons License https://creativecommons.org/licenses/by/4.0/). (**c**) 2D-bifurcation diagram showing the effects of the two mayor genetic risk factors for atopic dermatitis on the "dynamic phenotypes." (**e**) Long-term behavior of the barrier integrity of the four "dynamical phenotypes." (**f**) Barrier recovery time of the control vs- bistable (healthy branch) "dynamical phenotypes." Figures (**c**)–(**f**) adapted from https://doi.org/10.1016/j.jaci.2016.10.026 (published under the Creative Commons License https://creativecommons.org/licenses/by/4.0/)

Peptides (AMPs) [309], decreasing the strength of the negative feedback mediated by the immune responses (Fig. 3.21b). To systematically characterize the effects of different combinations of these two risk factors, in [123] we proposed a multi-scale model that comprises the fast control of the innate immune responses, and the slower tissue-level processes that are affected by these risk factors. The hybrid system of differential equations (see Sect. 2.9) is given by:

$$\frac{dP(t)}{dt} = P_{\text{env}} \frac{\kappa_P}{1 + \gamma_B B(t)} - \alpha_I R(t) P(t) - \delta_P P(t),$$

$$\frac{dB(t)}{dt} = \kappa_B \frac{1}{1 + \gamma_R R(t)} (1 - B(t)) - \delta_B K(t) B(t),$$

(3.3)

for the dynamics of the tissue-level variables $P(t)$ and $B(t)$, denoting the infiltrated pathogen load (mg/ml) and the strength of barrier integrity (relative to the maximum strength), respectively. The dynamics of $P(t)$ and $B(t)$ depend on the dynamics of the additional variables, $R(t)$ and $K(t)$, denoting the levels of activated immune receptors and active KLKs, respectively. We consider that the infiltrated pathogen load, $P$, increases by the penetration of environmental stress load, $P_{\text{env}}$, through the barrier, $B$. $P$ is eradicated by innate immune responses triggered by inflammation ($R$) and is also naturally degraded. The barrier production is described by phenomenological representation of its capacity to self-restore the nominal barrier function following its disruption, and is compromised by innate immune responses triggered by inflammation ($R$). We represent this inhibitory rate by the phenomenological term $1/(1 + X)$. The degradation of the barrier occurs as a result of desquamation mediated by active kallikreins, $K$.

To describe the activity of $R(t)$ and $K(t)$ in response to the dynamically changing infiltrated pathogen load $P$, we phenomenologically describe the mechanistically derived bifurcation diagrams (Fig. 3.19c) with a perfect switch (Fig. 3.21a), as:

$$(R(t), K(t)) = \begin{cases} (R_{\text{off}}, K_{\text{off}}), \\ \quad \text{if } (P(t) < P^-) \text{ or } \left\{ P(t) \in [P^-, P^+] \text{ and } R(t^-) = R_{\text{off}} \right\}, \\ (R_{\text{on}}, K_{\text{on}} = m_{\text{on}} P(t) - \beta_{\text{on}}), \\ \quad \text{if } (P(t) < P^-) \text{ or } \left\{ P(t) \in [P^-, P^+] \text{ and } R(t^-) = R_{\text{on}} \right\}, \end{cases}$$

(3.4)

where $t^-$ is a time slightly before the time $t$, as they change abruptly within hours [124, 502], in a much faster time-scale than for $P(t)$, $B(t)$, and $D(t)$, which change over weeks [202, 515].

Together, the coupling between Eqs. (3.3) and (3.4) comprises a hybrid system of algebraic-differential equations. To analyze the different qualitative behaviors that can result from sweeping the strengths of the tissue-level positive and negative feedbacks, we performed the *focal point analysis* described in Sect. 2.9 and schematically represented in Fig. 2.26. We found that different severities of the tissue-level risk factors can lead to four different qualitative dynamic behaviors (Fig. 3.21c, d):

1. In the absence of genetic risk factors, the system quickly recovers from environmental perturbations (in the form of a transient increase in $P_{env}$). After a transient decrease, the skin barrier returns to its nominal state; analogously, after a atopic dermatitis flare, the immune responses turn off ($R = R_{off}$). This *dynamical phenotype* corresponds to the healthy control.
2. When both genetic risk factors are present, the system converges to a unhealthy steady state, corresponding to a chronically decreased barrier integrity and persistent flares ($R = R_{on}$). This corresponds to the *unhealthy dynamical phenotype* in Fig. 3.21c, d.
3. Genetic defects leading to deregulated immune responses result in bistability, where either the healthy or unhealthy steady state is achieved depending on the initial conditions of the system (*bistable* dynamical phenotype). Interestingly, even when the environmental triggers are low enough for the system to remain in the healthy basin of attraction (Fig. 2.21), computational analysis of the model shows that the recovery time is significantly slower in the *bistability dynamical phenotype* as compared to the *control* dynamical phenotype (Fig. 3.21f). Mathematically, this observation can be related to the existence of the second stable (unhealthy) steady state, which is responsible to the *critical slowing down* of nonlinear dynamical systems that are close to a bifurcation [106]. Experimentally, this model prediction is consistent with the slower skin barrier recovery following tape stripping observed in non-lesional skin of patients suffering atopic dermatitis compared with healthy individuals [431], and in inflamed compared with non-inflamed human skin [203]. This model prediction is clinically relevant, since this "critical slowing down" of the skin barrier could be used to distinguish asymptomatic carriers of genetic risk factors that decrease the immune responses to infiltrated pathogens.
4. Genetic defects leading to high skin barrier permeability results in persistent oscillatory dynamics due to the switching of $R$ between $R_{on}$ and $R_{off}$ (*oscillation dynamic phenotype*).

Interestingly, the steady-state behaviors of the *bistability* (healthy branch) and dynamical phenotypes are clinically indistinguishable from the *control* dynamical phenotype (Fig. 3.21e). Although the *oscillation dynamical phenotype* shows a lower mean, it also has an increased variance, which makes it statistically hard to distinguish from the *control dynamical phenotype*. This long-term dynamical behavior of the oscillation phenotype is concordant with a slightly lower but more variable skin barrier integrity observed in mouse models for atopic dermatitis carrying mutations in the filaggrin gene ($flg^{-/-}$ [242] and $ft$ [143, 403]), compared to their *wild type (wt)* litter mates. Thus, according to these results, the presence of a single genetic risk factor is not *per se* associated to the development of clinically detectable atopic dermatitis symptoms. Rather, the development of severe symptoms of atopic dermatitis might require a *second hit*, for example, in the form of environmental insults. This observation is consistent with experimental observations, stating that animal models of atopic dermatitis with single genetic defects (e.g., *Stat3-ko* [122], *ft/ft* [403], *flg$^{-/-}$* [242]) require environmental triggers to develop clinically severe atopic dermatitis.

In what follows we present a mathematical model that reproduces this atopic dermatitis progression, from an asymptomatic stage to the development of the severe disease. The implications of these results in terms of prevention of severe symptoms are discussed.

## *Preventing Disease Progression*

Atopic dermatitis is a multi-stage disease, in which early and asymptomatic phases can progress to advanced atopic dermatitis characterized by severe symptoms that are difficult to treat. Although several genetic and environmental factors have been associated to development, the mechanisms through which different combinations of these factors contribute to the progression of atopic dermatitis had not been fully elucidated.

Recently, this question of atopic dermatitis progression and its prevention was addressed through mathematical modeling [122]. We extended the model of early phases of atopic dermatitis [123] previously exposed to incorporate key cellular and molecular players responsible for the severe symptoms of atopic dermatitis: the adaptive immune responses (activated Th2 cells) and its mediators (the transcription factor Gata3 and the dendritic cells) (Fig. 3.22). From previous experimental and clinical literature [143, 160, 190, 257, 296, 403, 437, 517] it was well known that Th2 cell activation and with it, the sharp increase in the severity of atopic dermatitis symptoms can be triggered by atopic dermatitis flares. This occurs through the cytokine-dependent activation of dendritic cells, which migrate to the lymph nodes where they increase the concentrations of the pro-inflammatory cytokine IL4, which triggers the expression of the Th2 cell differentiation marker Gata3. What remained to be elucidated, however, was:

- What are the quantitative features of atopic dermatitis flares necessary to trigger the aggravation of atopic dermatitis?
- How do risk factors affect the propensity to develop severe atopic dermatitis symptoms?
- Which role does the combination of genetic and environmental risk factors play in the establishment of severe atopic dermatitis symptoms?
- How can the prevention of the progression of atopic dermatitis be achieved? and:
- Do preventive strategies have to be tailored to the specific genetic background of the patient?

To answer these questions from a mathematical, systems biology approach, we incorporated the regulatory interactions for atopic dermatitis progression detailed above into the previous model by coupling to Eqs. (3.3) and (3.4) the dynamical equation describing the innate immune receptor-mediated migration of DC to the lymph nodes, i.e.:

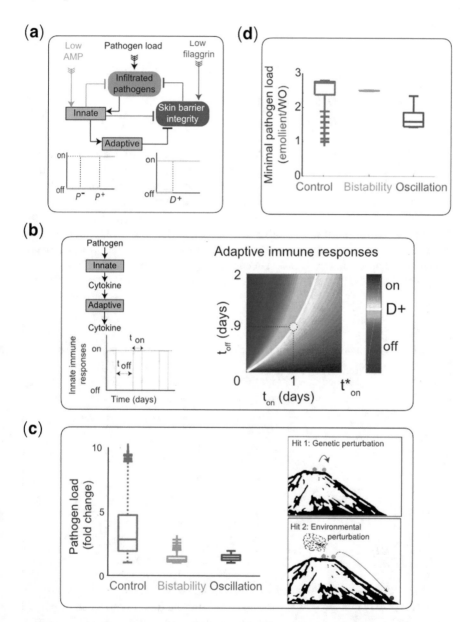

**Fig. 3.22** Mathematical model to understand and prevent the progression of Atopic Dermatitis. (**a**) "Double switch" model for the progression of atopic dermatitis. Progression occurs when adaptive immune responses are irreversibly turned on. (**b**) Mechanisms for progression of atopic dermatitis. Long-lasting or frequent atopic dermatitis flares irreversibly switch on the adaptive immune responses. (**c**) Genetic risk factors increase the susceptibility to develop severe atopic dermatitis symptoms by decreasing the minimal pathogen load required for triggering long-lasting or frequent atopic dermatitis flares. (**d**) Emollients effectively prevent the progression of AD by increasing the minimal pathogen load required for allergic sensitization of the skin. Figures adapted from [122] URL: https://doi.org/10.1016/j.jaci.2016.10.026, published under the Creative Commons License https://creativecommons.org/licenses/by/4.0/

$$\frac{dD(t)}{dt} = \kappa_D R(t) - \delta_D D(t), \tag{3.5}$$

as well as the irreversible switch-like expression of Gata3 in T cells:

$$G(t) = \begin{cases} G_{\text{off}}, & \text{if } (D(t) < D^+) \text{ and } G(t^-) = G_{\text{off}}, \\ G_{\text{on}}, & \text{if } (D(t) \geq D^-) \text{ or } \left\{ D(t) < D^+ \text{ and } G(t^-) = G_{\text{on}} \right\}, \end{cases} \tag{3.6}$$

To address the first question (What are the quantitative features of atopic dermatitis flares necessary to trigger Atopic Dermatitis aggravation?), we used expressions (3.5) and (3.6) to analytically determine which type of atopic dermatitis flares—i.e., on-and-off activations of the innate immune responses with specific on-periods (with duration $t_{\text{on}}$) and off-periods (with duration $t_{\text{off}}$) could trigger the irreversible activation of the adaptive immune responses. From (3.5) and (3.6) we could derive analytical expressions for the critical values of $t_{\text{on}}^*$ and $t_{\text{off}}^*$ above/below which atopic dermatitis progression occurs, as shown in Box 3.2.

**Box 3.2.** Analytical derivation of $t_{\text{on}}^*$ and $t_{\text{off}}^*$

This mathematical derivation is taken from the Supplementary Information provided in [122] URL: https://doi.org/10.1016/j.jaci.2016.10.026, published under the Creative Commons License https://creativecommons.org/licenses/by/4.0/.

The solution of (3.5),

$$\frac{D(t)}{dt} = \kappa_{DC} R(t) - \delta_D D(t),$$

is described by

$$D(t) = e^{-\delta_D(t-t_0)} D(t_0) + \int_{t_0}^{t} e^{-\delta_D(t-\tau)} \kappa_{DC} R(\tau) d\tau, \tag{3.7}$$

where the integral is defined over each time segment, on which $R(t)$ is continuously defined, either by $R(t) = R_{\text{on}}$ for the duration of a flare time, $t_{\text{on}}$, or by $R(t) = R_{\text{off}} = 0$ for the duration of a relaxation time, $t_{\text{off}}$. Note that the steady-state value, $D_{\text{ss}}$, of $D(t)$ while $R(t) = R_{\text{on}}$ is obtained by $D_{\text{ss}} = \frac{\kappa_{DC} R_{\text{on}}}{\delta_D}$. The period of the $R$-spike is denoted by $T = t_{\text{on}} + t_{\text{off}}$.
To determine the dynamics of $D(t)$, we derive $D(t_k)$ and $D(T_k)$ ($k = 1, 2, \ldots$), where $t_k$ and $T_k$ denote the time when the $k$-th spike of $R(t) = R_{\text{on}}$ starts and the time when the $k$-th spike ends, respectively. We define $t_1 = 0$

(continued)

**Box 3.2.** (continued)

and $D(t_1) = 0$. $D(t)$ decreases during $T_k \leq t \leq t_{k+1}$ when $R(t) = R_{\text{off}} = 0$ and reaches $D(t_{k+1}) = e^{-\delta_D t_{\text{off}}} D(T_k)$, whereas it increases during $t_k \leq t \leq T_k$ with $R(t) = R_{\text{on}}$ and achieves

$$D(T_k) = e^{-\delta_D t_{\text{on}}} D(t_k) + \kappa_{DC} R_{\text{on}} \int_{t_k}^{T_k} e^{-\delta_D(T_k - \tau)} d\tau$$

$$= e^{-\delta_D t_{\text{on}}} D(t_k) + \kappa_{DC} R_{\text{on}} e^{-\delta_D T_k} \int_{t_k}^{T_k} e^{\delta_D \tau} d\tau$$

$$= e^{-\delta_D t_{\text{on}}} D(t_k) + D_{ss} (1 - e^{-\delta_D t_{\text{on}}}). \tag{3.8}$$

Since $D(t_1) = 0$ for $t_1 = 0$, we have $D(T_1) = D_{ss} (1 - e^{-\delta_D t_{\text{on}}})$ and

$$D(T_k) = e^{-\delta_D t_{\text{on}}} e^{-\delta_D t_{\text{off}}} D(T_{k-1}) + D_{ss} (1 - e^{-\delta_D t_{\text{on}}})$$
$$= e^{-\delta_D T} D(T_{k-1}) + D_{ss} (1 - e^{-\delta_D t_{\text{on}}}), \quad k = 2, 3, \ldots.$$

Therefore, $D(T_k)$ is described as

$$D(T_k) = D_{ss} \left(1 - e^{-\delta_D t_{\text{on}}}\right) \sum_{i=0}^{k-1} e^{-i\delta_D T}, \tag{3.9}$$

which converges to

$$D(T_\infty) = \lim_{k \to \infty} D(T_k) = D_{ss}(1 - e^{-\delta_D t_{\text{on}}}) \frac{1}{1 - e^{-\delta_D T}}. \tag{3.10}$$

The minimum flare time, $t_{\text{on}}^*$, for a single pulse of $R(t) = R_{\text{on}}$ to trigger systemic Th2 sensitization is analytically obtained from the corresponding solution

$$D(t) = \int_0^t e^{-\delta_D(t-\tau)} \kappa_{DC} R(\tau) d\tau = D_{ss}(1 - e^{-\delta_D t})$$

of Eq. (3.7). Solving $D(t_{\text{on}}^*) = D^+$ leads to

$$t_{\text{on}}^* = -\frac{\ln\left(1 - \frac{D^+}{D_{ss}}\right)}{\delta_D}. \tag{3.11}$$

The minimum relaxation time, $t_{\text{off}}^*$, for a periodic $R$-spike with a fixed $t_{\text{on}}$ to trigger systemic Th2 sensitization is analytically obtained by solving $D(T_\infty) = D^+$ as

(continued)

Box 3.2.    (continued)

$$t_{\text{off}}^* = -\left[ t_{\text{on}} + \frac{1}{\delta_D} \ln\left\{ 1 - \frac{D_{ss}}{D^+}(1 - e^{-\delta_D t_{\text{on}}}) \right\} \right]. \qquad (3.12)$$

Note that the solution in Eq. (3.12) exists if $1 - \dfrac{D_{ss}}{D^+}(1 - e^{-\delta_D t_{\text{on}}}) > 0$ and

$t_{\text{on}} + \dfrac{1}{\delta_D} \ln\left\{ 1 - \dfrac{D_{ss}}{D^+}(1 - e^{-\delta_D t_{\text{on}}}) \right\} < 0$. These conditions are equivalent to

$$t_{\text{on}} < -\frac{1}{\delta_D} \ln\left( 1 - \frac{D^+}{D_{ss}} \right) = t_{\text{on}}^* \quad \text{and} \quad D^+ < D_{ss}.$$

These results, represented also in Fig. 3.22b, show that only long-lasting (i.e., with $t_{\text{on}} > t_{\text{on}}^*$) or frequent (i.e., with $t_{\text{off}} < t_{\text{off}}^*$) flares can induce the Th2 cell activation.

Next, we investigate how the two most frequent genetic risk factors for atopic dermatitis, i.e.:

- *low filaggrin expression*, augmenting the barrier permeability (parameter $\kappa_P$ in the model)
  and:
- *low immune responses*, decreasing the efficacy of the innate immune responses to eradicate pathogens (parameter $\alpha_I$ in the model) as well as increased pathogen load (corresponding to the most common *environmental* risk factor for atopic dermatitis)

synergistically affect the progression of atopic dermatitis (Fig. 3.22c).

For this, we considered the three *virtual patient cohorts*, corresponding to three of the four regions of the bifurcation diagram in Fig. 3.21c:

- The "Control" virtual patient cohort.
- The "Bistability" virtual patient cohort.
  and:
- The "Oscillation" virtual patient cohort.

Recall that single genetic risk factors (high barrier permeability and low immune responses) often give rise to the asymptomatic phenotypes ("oscillation" and "bistability"), which at steady state cannot be distinguished from the "control" virtual patient cohort (Fig. 3.21e). This means that these genetic alterations are not enough for the full development of a severe atopic dermatitis phenotype. So:

- What else is required?
- What is the role of environmental risk factors on atopic dermatitis progression, in different genetic backgrounds?

In other words:

- Which role does the combination of genetic and environmental risk factors play in the establishment of severe atopic dermatitis symptoms?

To answer these questions, for each of these virtual patient cohorts, we calculated numerically the *minimal pathogen load required to trigger allergic sensitization*; i.e., the minimal pathogen load required to trigger long-lasting (i.e., with $t_{on} > t_{on}^*$) or frequent (i.e., with $t_{off} < t_{off}^*$) atopic dermatitis flares, as previously analytically determined (see Box 3.2 and Fig. 3.22c). The results of these investigations are shown in Fig. 3.22c. For the "control" virtual patient cohort with no genetic risk factors, a high pathogen load is required to trigger allergic sensitization. Genetic risk factors, initially appearing as an asymptomatic phenotype, drastically decrease this minimal amount of pathogen load, hence increasing the susceptibility of (asymptomatic) genetic risk factor carriers to develop severe symptoms of atopic dermatitis in response to environmental challenges. This synergistic effect of genetic and environmental factors can be seen as a "*two-hit-process*" for atopic dermatitis progression: The first genetic hit results in an asymptomatic phenotype that is phenotypically indistinguishable from the control phenotype but much more vulnerable, and a second hit, in the form of environmental fluctuations causing natural variations in pathogen loads.

We can visualize this -two-hit-process with the following analogy:

> A genetic perturbation acts as on the position ("phenotype") of a ball on a mountain (location in the epigenetic landscape) by pushing it to the edge of a mountain top - it's still on the same level (phenotype) as the control genotype, but more vulnerable to environmental perturbations: it is much easier for a "genotype on the edge" to roll down the hill and develop severe symptoms! (Fig. 3.22).

Finally, the results of this investigation were used to evaluate the effects of emollients on the progression of atopic dermatitis. Specifically, we aimed at answering the two remaining questions, i.e.:

- How can atopic dermatitis progression be prevented?
and:
- Do preventive strategies have to be tailored to the specific genetic background of the patient?

For this, we first mathematically represented emollient treatments by adding a constant term $+E$ to $dB(t)/dt$ in (3.3). Recent clinical trials [206, 420] suggested that emollients could effectively prevent the development of severe atopic dermatitis. However, due to the large variation in the patient cohort considered in this investigation and the lack of mechanistic understanding of the effects of emollients on a heterogeneous group of patients, the result of these trials were not conclusive. To determine if application of emollients could effectively prevent the progression of atopic dermatitis from a mild to a severe phenotype, regardless of the genetic background, we evaluated the susceptibility of the different virtual patient cohorts to develop severe atopic dermatitis symptoms in response to pathogenic challenges.

Our results show that, regardless of the genetic background, emollient application decreases the susceptibility to develop severe atopic dermatitis symptoms by decreasing the minimal pathogen load required for atopic dermatitis progression (Fig. 3.22d).

Together, the results presented in this section (Fig. 3.22) show that mathematical modeling can be used to:

- Uncover the mechanisms of complex disease progression.
- Characterize the effects of different risk factor combinations on the pathogenic process.
  and:
- Systematically and quantitatively evaluate the benefits of preventive treatment strategies.

We can at this level conclude our quantitative example with some final comments.

## Concluding Remarks

In this section we showcased how multi-scale, mechanistic ordinary differential equations models can be used to gain understanding of the mechanisms underlying the onset and progression of a complex epithelial tissue disease: atopic dermatitis. We also showed how such a quantitative framework can be used to improve the early detection strategies to identify the susceptible but asymptomatic patient cohorts that might benefit from preventive treatments, and to systematically test the efficacy of preventive treatment regimens on the pathogenic progression, and whether this efficacy is affected by the genetic background of the patients.

Furthermore, such dynamical, mechanistic, integrative, and quantitative systems-level representations of a disease process can be used for the design of optimized and -patient-specific treatment strategies that effectively revert the symptoms using the minimal amount of pharmacological treatment. For example, in [88], Christodoulides et al. used optimal control theory to minimize the use of corticosteroids in pro-active treatments [441] of severe atopic dermatitis forms. In [438], Tanaka et al. used a bifurcation analysis approach to systematically design -patient-specific optimal treatments for atopic dermatitis.

Minimizing the duration and magnitude of pharmacological treatments used for the reversion of severe symptoms is a clinically relevant task, since many pharmacological treatments have associated negative side effects. For example, reducing epithelial inflammation with corticosteroids might lead to tissue atrophy [407]. Analogously, immoderate use of antibiotics to treat epithelial infection is associated to the development of antibiotics resistant bacterial strains [352, 377, 408].

We hope that with this example we could show how mathematical models of complex diseases can be used to gain an integrative, quantitative, and dynamical systems-level understanding of complex diseases, such as atopic dermatitis, to improve their early detection, prevention, and treatment strategies (see Fig. 3.23).

**Fig. 3.23** Mathematical models of complex epithelial tissue diseases, such as atopic dermatitis and carcinomas, provide a formal, quantitative, and systems-level framework to analyze the causes for onset and progression and to improve the treatment strategies for disease prevention and reversión

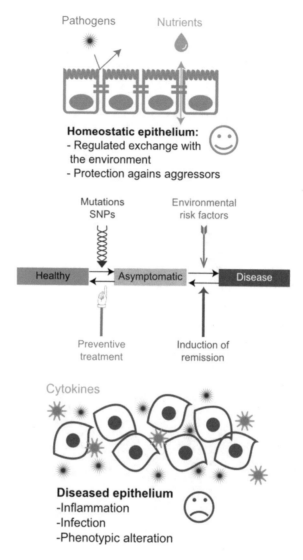

We would like to encourage the use of mathematical models of complex diseases, since, as shown here, they have the potential to reduce the burden of diseases and thus contribute to the health and well-being of the population.

# Conclusions and Perspectives

In this volume we have presented basic concepts and formal/computational approaches of systems biology using a bottom-up approach as applied to complex chronic degenerative diseases. To this end we have emphasized the use of Boolean network dynamical modeling to understand the systemic mechanisms underlying the emergence of different cell types under health and disease, but also the extension of such models to continuous approximations to study quantitative dynamics and the role of specific network nodes in response to possible microenvironmental or physicochemical cues. In addition to complex intracellular regulatory networks, signal transduction pathways couple to these networks and mediate cellular feedback to physicochemical and environmental factors also under normal and altered conditions.

Complex intracellular networks attain stable or attractor states that correlate to the gene/protein expression/activation configurations that are observed in different cell types and that are coherent with the restrictions imposed by network interactions. Such intracellular networks also give rise to multi-dimensional quasi-potentials that may be considered the mathematical or formal representations of the epigenetic landscape proposed by Conrad Hal Waddington. In this book we have also described our deterministic and stochastic approaches to formal and quantitative analyses of the epigenetic landscape that restricts time-ordered and spatial patterns of cell transitions and dynamics. These approaches are useful to further understand the systemic mechanisms that may underlie the conserved patterns of progression during complex diseases. We illustrated these using the case of epithelial cancer, which is the most common type of cancer, characterized by the progression of normal epithelial tissues to tissues affected by chronic inflammation and then to mesenchymal neoplasias, regardless of the type of epithelium. We summarize here our findings of a regulatory network module that recovers as attractor states the configurations corresponding to epithelial, inflamed, and mesenchymal cells and show that this module also yields the time-ordered transitions observed in the spontaneous immortalization of epithelial cell cultures and could provide a

© Springer International Publishing AG, part of Springer Nature 2018
M. E. Álvarez-Buylla Roces et al., *Modeling Methods for Medical Systems Biology*,
Advances in Experimental Medicine and Biology 1069,
https://doi.org/10.1007/978-3-319-89354-9

basic framework to further understanding the systems-level mechanisms underlying cancer emergence and progression dynamics.

In the second case study, we explored how the plasticity of CD4+ T cells is modulated by microenvironmental (pro-inflammatory cytokines) and lifestyle (diet) factors. For this, we integrated in a mathematical model both the key transcriptional regulators, signaling molecules, and extracellular signals (cytokines) that underlie the patterns of CD4+ T-cell differentiation and plasticity. Moreover, we were able to exemplify how such cellular networks may be linked to alterations associated to various types of complex diseases. Particularly, we have explored the system-level mechanisms by which hyperinsulinemia causes chronic inflammation and the types of couplings that may occur between altered metabolites or hormones and CD4+ T-cell differentiation under various diseases that imply alterations of the immune system. Specific and more complete feedback mechanisms among the uncovered network and metabolic network modules awaits further analyses. Using this case we also exemplify an approach that we have proposed to analyze the re-patterning of the epigenetic landscape in the context of a continuous network model in which the impact of continuous alterations in the decay rates of particular components or nodes can be simulated. We have particularly shown how different cytokines can alter the immune system by modifying CD4+ T-cell differentiation and plasticity dynamics. The network proposed in this case, together with other similar studies, should become an important modeling framework to explore the link between inflammation, the immune system, and several diseases.

A third case study illustrates a scenario in which quantitative, mechanistic models based on kinetic interactions are used to capture the onset and progression of atopic dermatitis. For this, multi-scale, mechanistic kinetic models were used to capture the dynamic interplay between coupled biochemical and tissue-level processes that underlie epidermal function in health and disease. We also show how the model predictions are validated with clinical and experimental *in vivo data*. Further, we illustrate how we use such models to design therapeutic strategies to prevent or revert severe symptoms, using control theory approaches.

As novel single-cell Omics approaches, especially deep sequencing, transcriptomic, proteomics, and epigenomic ones, are developed, we expect rapid progress in research that links such empirical top-down description of single cell behavior with the types of dynamic, bottom-up approaches that we presented here. For example, it would be interesting to explore to what extent the networks inferred with top-down approaches can be dynamically analyzed using the tools presented here. Specifically, it would be interesting to assess whether the empirically described individual single-cell profiles along the temporal differentiation trajectories that are being studied can be mathematically reproduced by the transient and attractor states of the underlying multistable networks.

The integrative approach put forward here has also been very useful to identify holes in the empirical evidence accumulated up to now. As mathematical models of biological systems integrate experimental data, their analysis and predictions can be used to guide further experimental approaches, suggesting important missing components or interactions of the systems involved.

The type of approach that we and other research groups have followed is also contributing to the generation of a repertoire of relevant gene regulatory network modules that enable uncovering patterns shared by different types of network modules, and eventually contribute to understanding how network structure relates to the function and dynamics of such regulatory modules. For example, the modules that have been uncovered up to now seem to be quite robust to many different types of structural and functional perturbations. Further, the existence of multiple attractors in these models seems to result from specific combinations of positive and negative feedback loops. Also, positive feedback loops seem to be important for making networks robust to attacks to highly connected nodes [31, 141]. Other research groups have also started to analyze the prevalence of certain regulatory motifs and their function in the context of the whole networks (for example, [7]).

As more sophisticated time-lapse, single-cell, and life-imaging approaches are developed, additional data will become available to better calibrate and experimentally validate systems-level models. Multilevel modeling will be particularly relevant and useful as such data becomes available. For example, a rapidly changing research area is addressing the role of mechanic and elastic fields in morphogenesis and in disease emergence and progression. Novel approaches that couple such physical forces with the intracellular complex regulatory network modules, as the ones presented here, should be further pursued.

We believe that the approach presented in this book is a first building block in systems biology modeling. This and further formal and computational approaches are necessary for the mechanistic understanding of the role of genetic versus environmental components in disease emergence and progression. Since systems biology approaches can be used to improve the strategies for early detection, diagnosis, prevention, and treatment of complex diseases, we predict that their use will significantly improve health care programs by establishing prospective analyses grounded on a combination of mathematical system biology models as well as experimental and clinical data.

# Bibliography

1. Aberg, K. M., Man, M. Q., Gallo, R. L., Ganz, T., Crumrine, D., Brown, B. E., et al. (2008). Co-regulation and interdependence of the mammalian epidermal permeability and antimicrobial barriers. *Journal of Investigative Dermatology, 128*(4), 917–925.
2. Abou-Jaoudé, W., Monteiro, P. T., Naldi, A., Grandclaudon, M., Soumelis, V., Chaouiya, C., et al. (2015). Model checking to assess T-helper cell plasticity. *Frontiers in Bioengineering and Biotechnology, 2*, 86.
3. Adler, M., & Alon, U. (2018). Fold-change detection in biological systems. *Current Opinion in Systems Biology, 8*, 81–89.
4. Ahn, A. C., Tewari, M., Poon, C. S., & Phillips, R. S. (2006). The limits of reductionism in medicine: Could systems biology offer an alternative?. *PLoS Medicine, 3*(6), e208.
5. Alberch, P. (1991). From genes to phenotype: Dynamical systems and evolvability. *Genetica, 84*, 5–11.
6. Albert, I., Thakar, J., Li, S., Zhang, R., & Albert, R. (2008). Boolean network simulations for life scientists. *Source Code for Biology and Medicine, 3*(1), 16.
7. Albert, R. (2007). Network inference, analysis, and modeling in systems biology. *The Plant Cell Online, 19*(11), 3327–3338.
8. Albert, R., & Barabási, A. L. (2002). Statistical mechanics of complex networks. *Reviews of Modern Physics, 74*(1), 47.
9. Albert, R., & Othmer, H. G. (2003). The topology of the regulatory interactions predicts the expression pattern of the segment polarity genes in Drosophila melanogaster. *Journal of Theoretical Biology, 223*(1), 1–18.
10. Aldana, M., Balleza, E., Kauffman, S., & Resendiz, O. (2007). Robustness and evolvability in genetic regulatory networks. *Journal of Theoretical Biology, 245*(3), 433–448.
11. Al-Hashmi, S., Ekanayake, M., & Martin, C. (2009). Type II diabetes and obesity: A control theoretic model. *Emergent Problems in Nonlinear Systems and Control, 393*, 1–19.
12. Aliahmadi, E., Gramlich, R., Grützkau, A., Hitzler, M., Krüger, M., Baumgrass, R., et al. (2009). TLR2-activated human Langerhans cells promote Th17 polarization via IL-1$\beta$, TGF-$\beta$ and IL-23. *European Journal of Immunology, 39*(5), 1221–1230.
13. Allam, M. F., & Arjona, M. O. (2013). Health promotion or pharmacological treatment for chronic diseases? *Journal of Preventive Medicine and Hygiene, 54*(1), 11.
14. Alon, U., Surette, M. G., Barkai, N., & Leibler, S. (1999). Robustness in bacterial chemotaxis. *Nature, 397*(6715), 168–171.
15. Altrock, P. M., Liu, L. L., & Michor, F. (2015). The mathematics of cancer: Integrating quantitative models. *Nature Reviews Cancer, 15*(12), 730–745.

© Springer International Publishing AG, part of Springer Nature 2018
M. E. Álvarez-Buylla Roces et al., *Modeling Methods for Medical Systems Biology*,
Advances in Experimental Medicine and Biology 1069,
https://doi.org/10.1007/978-3-319-89354-9

16. Álvarez-Buylla, E. R., Azpeitia, E., Barrio, R., Benítez, M., & Padilla-Longoria, P. (2010). From ABC genes to regulatory networks, epigenetic landscapes and flower morphogenesis: Making biological sense of theoretical approaches. *Seminars in Cell & Developmental Biology, 21*(1), 108–117.

17. Álvarez-Buylla, E. R., Benítez, M., Davila, E. B., Chaos, A., Espinosa-Soto, C., & Padilla-Longoria, P. (2007). Gene regulatory network models for plant development. *Current Opinion in Plant Biology, 10*(1), 83–91.

18. Álvarez-Buylla, E. R., Chaos, Á., Aldana, M., Benítez, M., Cortes-Poza, Y., Espinosa-Soto, C., et al. (2008). Floral morphogenesis: Stochastic explorations of a gene network epigenetic landscape. *PLoS One, 3*(11), e3626.

19. Álvarez-Buylla, E. R., Dávila-Velderrain, J., & Martínez-García, J. C. (2016). Systems biology approaches to development beyond bioinformatics: Nonlinear mechanistic models using plant systems. *BioScience, 66*(5), 371–383.

20. Anand, P., Kunnumakara, A. B., Sundaram, C., Harikumar, K. B., Tharakan, S. T., Lai, O. S., et al. (2008). Cancer is a preventable disease that requires major lifestyle changes. *Pharmaceutical Research, 25*(9), 2097–2116.

21. Anderson, D. F., & Kurtz, T. G. (2015). *Stochastic analysis of biochemical systems* (Vol. 1). Berlin: Springer.

22. Angeli, D., Ferrell, J. E., & Sontag, E. D. (2004). Detection of multistability, bifurcations, and hysteresis in a large class of biological positive-feedback systems. *Proceedings of the National Academy of Sciences of the United States of America, 101*(7), 1822–1827.

23. Antebi, Y. E., Nandagopal, N., & Elowitz, M. B. (2017). An operational view of intercellular signaling pathways. *Current Opinion in Systems Biology, 1*, 16–24.

24. Ao, P., Galas, D., Hood, L., & Zhu, X. (2008). Cancer as robust intrinsic state of endogenous molecular-cellular network shaped by evolution. *Medical Hypotheses, 70*(3), 678–684.

25. Arellano, G., Argil, J., Azpeitia, E., Benítez, M., Carrillo, M., Góngora, P., et al. (2011). "Antelope": A hybrid-logic model checker for branching-time Boolean GRN analysis. *BMC Bioinformatics, 12*(1), 490.

26. Arendt, D., Musser, J. M., Baker, C. V., Bergman, A., Cepko, C., Erwin, D. H., et al. (2016). The origin and evolution of cell types. *Nature Reviews Genetics, 17*(12), 744–757.

27. Arroyo, A. G., & Iruela-Arispe, M. L. (2010). Extracellular matrix, inflammation, and the angiogenic response. *Cardiovascular Research, 86*(2), 226–235.

28. Azpeitia, E., Benítez, M., Padilla-Longoria, P., Espinosa-Soto, C., & Álvarez-Buylla, E. R. (2011). Dynamic network-based epistasis analysis: Boolean examples. *Frontiers in Plant Science, 2*, 92.

29. Azpeitia, E., Benítez, M., Vega, I., Villarreal, C., & Álvarez-buylla, E. R. (2010). Single-cell and coupled GRN models of cell patterning in the Arabidopsis thaliana root stem cell niche. *BMC Systems Biology, 4*, 134.

30. Azpeitia, E., Davila-Velderrain, J., Villarreal, C., & Álvarez-Buylla, E. R. (2014). Gene regulatory network models for floral organ determination. In *Flower development: Methods and protocols* (pp. 441–469). New York: Humana Press.

31. Azpeitia, E., Muñoz, S., González-Tokman, D., Martínez-Sánchez, M. E., Weinstein, N., Naldi, A., et al. (2017). The combination of the functionalities of feedback circuits is determinant for the attractors' number and size in pathway-like Boolean networks. *Scientific Reports, 7*, 42023.

32. Azpeitia, E., Weinstein, N., Benítez, M., Mendoza, L., & Álvarez-Buylla, E. R. (2013). Finding missing interactions of the Arabidopsis thaliana root stem cell niche gene regulatory network. *Frontiers in Plant Science, 4*, 110.

33. Bäckman, A., Ny, A., Edlund, M., Ekholm, E., Hammarström, B. E., Törnell, J., et al. (2002). Epidermal overexpression of stratum corneum chymotryptic enzyme in mice: A model for chronic itchy dermatitis. *Journal of Investigative Dermatology, 118*(3), 444–449.

34. Bak, P., & Paczuski, M. (1995). Complexity, contingency, and criticality. *Proceedings of the National Academy of Sciences of the United States of America, 92*(15), 6689–6696.

35. Baker, B. S. (2006). The role of microorganisms in atopic dermatitis. *Clinical & Experimental Immunology, 144*(1), 1–9.
36. Balázsi, G., van Oudenaarden, A., & Collins, J. J. (2011). Cellular decision making and biological noise: From microbes to mammals. *Cell, 144*(6), 910–925.
37. Baldwin, G., Bayer, T., Dickinson, R., Ellis, T., Freemont, P. S., Kitney, R. I., et al. (2015). *Synthetic biology—A primer*. Singapore: World Scientific. Revised edition.
38. Balkwill, F., & Mantovani, A. (2001). Inflammation and cancer: Back to Virchow? *The Lancet, 357*(9255), 539–545.
39. Balleza, E., Alvarez-Buylla, E. R., Chaos, A., Kauffman, S., Shmulevich, I., & Aldana, M. (2008). Critical dynamics in genetic regulatory networks: Examples from four kingdoms. *PLoS One, 3*(6), e2456.
40. Bargaje, R., Trachana, K., Shelton, M. N., McGinnis, C. S., Zhou, J. X., Chadick, C., et al. (2017). Cell population structure prior to bifurcation predicts efficiency of directed differentiation in human induced pluripotent cells. *Proceedings of the National Academy of Sciences of the United States of America, 114*(9), 2271–2276.
41. Barrio, R. A. (2008). Turing systems: A general model for complex patterns in nature. In *Physics of emergence and organization* (pp. 267–296). Singapore: World Scientific. ISBN: 13 978-981-277-994-6, ISBN: 10 981-277-994-9.
42. Barrio, R. A., Hernandez-Machado, A., Varea, C., Romero-Arias, J. R., & Álvarez-Buylla, E. (2010). Flower development as an interplay between dynamical physical fields and genetic networks. *PLoS One, 5*(10), e13523.
43. Barrio, R. A., Romero-Arias, J. R., Noguez, M. A., Azpeitia, E., Ortiz-Gutiérrez, E., Hernández-Hernández, V., et al. (2013). Cell patterns emerge from coupled chemical and physical fields with cell proliferation dynamics: The Arabidopsis thaliana root as a study system. *PLoS Computational Biology, 9*(5), e1003026.
44. Batchelor, E., Loewer, A., Mock, C., & Lahav, G. (2011). Stimulus-dependent dynamics of p53 in single cells. *Molecular Systems Biology, 7*(1), 488.
45. Batchelor, E., Mock, C. S., Bhan, I., Loewer, A., & Lahav, G. (2008). Recurrent initiation: A mechanism for triggering p53 pulses in response to DNA damage. *Molecular Cell, 30*(3), 277–289.
46. Battula, V. L., Evans, K. W., Hollier, B. G., Shi, Y., Marini, F. C., Ayyanan, A., et al. (2010). Epithelial-mesenchymal transition-derived cells exhibit multilineage differentiation potential similar to mesenchymal stem cells. *Stem Cells, 28*(8), 1435–1445.
47. Becskei, A., & Serrano, L. (2000). Engineering stability in gene networks by autoregulation. *Nature, 405*(6786), 590–593.
48. Ben-Haim, N., Lu, C., Guzman-Ayala, M., Pescatore, L., Mesnard, D., Bischofberger, M., et al. (2006). The nodal precursor acting via activin receptors induces mesoderm by maintaining a source of its convertases and BMP4. *Developmental Cell, 11*(3), 313–323.
49. Benítez, M., Espinosa-Soto, C., Padilla-Longoria, P., & Álvarez-Buylla, E. R. (2008). Interlinked nonlinear subnetworks underlie the formation of robust cellular patterns in Arabidopsis epidermis: A dynamic spatial model. *BMC Systems Biology, 2*(1), 98.
50. Benítez, M., Espinosa-Soto, C., Padilla-Longoria, P., Díaz, J., & Álvarez-Buylla, E. R. (2007). Equivalent genetic regulatory networks in different contexts recover contrasting spatial cell patterns that resemble those in Arabidopsis root and leaf epidermis: A dynamic model. *The International Journal of Developmental Biology, 51*, 139–155.
51. Ben-Porath, I., Thomson, M. W., Carey, V. J., Ge, R., Bell, G. W., Regev, A., et al. (2008). An embryonic stem cell-like gene expression signature in poorly differentiated aggressive human tumors. *Nature Genetics, 40*(5), 499–507.
52. Bernard, F. X., Morel, F., Camus, M., Pedretti, N., Barrault, C., Garnier, J., et al. (2012). Keratinocytes under fire of proinflammatory cytokines: Bona fide innate immune cells involved in the physiopathology of chronic atopic dermatitis and psoriasis. *Journal of Allergy, 2012*, 10. Article ID 718725.
53. Bevilacqua, C., & Ducos, B. (2017). Laser microdissection: A powerful tool for genomics at cell level. *Molecular Aspects of Medicine, 59*, 5–27.

54. Blaser, M. J. (2017). The theory of disappearing microbiota and the epidemics of chronic diseases. *Nature Reviews Immunology, 17*(8), 461–463.
55. Bluestone, J. A., Mackay, C. R., O'shea, J. J., & Stockinger, B. (2009). The functional plasticity of T cell subsets. *Nature Reviews Immunology, 9*(11), 811–816.
56. Bonelli, M., Shih, H. Y., Hirahara, K., Singelton, K., Laurence, A., Poholek, A., et al. (2014). Helper T cell plasticity: Impact of extrinsic and intrinsic signals on transcriptomes and epigenomes. In *Transcriptional control of lineage differentiation in immune cells* (pp. 279–326). Berlin: Springer International Publishing.
57. Bousquet, J., Anto, J. M., Sterk, P. J., Adcock, I. M., Chung, K. F., Roca, J., et al. (2011). Systems medicine and integrated care to combat chronic noncommunicable diseases. *Genome Medicine, 3*(7), 43.
58. Boveri, T. (1914). *The origin of malignant tumors* (M. Boveri (1929) Baillière, Tindall & Cox, Trans.). Baltimore: The Williams and Wilkins Company.
59. Brady, T., Roth, S. L., Malani, N., Wang, G. P., Berry, C. C., Leboulch, P., et al. (2011). A method to sequence and quantify DNA integration for monitoring outcome in gene therapy. *Nucleic Acids Research, 39*(11), e72.
60. Brattsand, M., Stefansson, K., Lundh, C., Haasum, Y., & Egelrud, T. (2005). A proteolytic cascade of kallikreins in the stratum corneum. *Journal of Investigative Dermatology, 124*(1), 198–203.
61. Bratus, A., Samokhin, I., Yegorov, I., & Yurchenko, D. (2017). Maximization of viability time in a mathematical model of cancer therapy. *Mathematical Biosciences, 294*, 110–119.
62. Breitfeld, D., Ohl, L., Kremmer, E., Ellwart, J., Sallusto, F., Lipp, M., et al. (2000). Follicular B helper T cells express CXC chemokine receptor 5, localize to B cell follicles, and support immunoglobulin production. *Journal of Experimental Medicine, 192*(11), 1545–1552.
63. Bressloff, P. C. (2014). *Stochastic processes in cell biology* (Vol. 41). New York: Springer.
64. Briggs, G. E., & Haldane, J. B. S. (1925). A note on the kinetics of enzyme action. *Biochemical Journal, 19*(2), 338.
65. Briot, A., Deraison, C., Lacroix, M., Bonnart, C., Robin, A., Besson, C., et al. (2009). Kallikrein 5 induces atopic dermatitis-like lesions through PAR2-mediated thymic stromal lymphopoietin expression in Netherton syndrome. *Journal of Experimental Medicine, 206*(5), 1135–1147.
66. Brown, S. J., Kroboth, K., Sandilands, A., Campbell, L. E., Pohler, E., Kezic, S., et al. (2012). Intragenic copy number variation within filaggrin contributes to the risk of atopic dermatitis with a dose-dependent effect. *Journal of Investigative Dermatology, 132*(1), 98–104.
67. Bruggeman, F. J., & Westerhoff, H. V. (2007). The nature of systems biology. *Trends in Microbiology, 15*(1), 45–50.
68. Buddenkotte, J., Stroh, C., Engels, I. H., Moormann, C., Shpacovitch, V. M., Seeliger, S., et al. (2005). Agonists of proteinase-activated receptor-2 stimulate upregulation of intercellular cell adhesion molecule-1 in primary human keratinocytes via activation of NF-κB. *Journal of Investigative Dermatology, 124*(1), 38–45.
69. Çağatay, T., Turcotte, M., Elowitz, M. B., Garcia-Ojalvo, J., & Süel, G. M. (2009). Architecture-dependent noise discriminates functionally analogous differentiation circuits. *Cell, 139*(3), 512–522.
70. Caligaris, C., Vázquez-Victorio, G., Sosa-Garrocho, M., Ríos-López, D. G., Marín-Hernández, A., & Macías-Silva, M. (2015). Actin-cytoskeleton polymerization differentially controls the stability of Ski and SnoN co-repressors in normal but not in transformed hepatocytes. *Biochimica et Biophysica Acta (BBA)-General Subjects, 1850*(9), 1832–1841.
71. Calzone, L., Fages, F., & Soliman, S. (2006). BIOCHAM: An environment for modeling biological systems and formalizing experimental knowledge. *Bioinformatics, 22*(14), 1805–1807.
72. Campbell, C., & Albert, R. (2014). Stabilization of perturbed Boolean network attractors through compensatory interactions. *BMC Systems Biology, 8*(1), 53.
73. Campisi, J., Andersen, J. K., Kapahi, P., & Melov, S. (2011). Cellular senescence: A link between cancer and age-related degenerative disease? *Seminars in Cancer Biology, 21*(6), 354–359.

74. Cannavó, F. (2012). Sensitivity analysis for volcanic source modeling quality assessment and model selection. *Computers & Geosciences, 44*, 52–59.
75. Cannons, J. L., Lu, K. T., & Schwartzberg, P. L. (2013). T follicular helper cell diversity and plasticity. *Trends in Immunology, 34*(5), 200–207.
76. Caubet, C., Jonca, N., Brattsand, M., Guerrin, M., Bernard, D., Schmidt, R., et al. (2004). Degradation of corneodesmosome proteins by two serine proteases of the kallikrein family, SCTE/KLK5/hK5 and SCCE/KLK7/hK7. *Journal of Investigative Dermatology, 122*(5), 1235–1244.
77. Cellier, F. E. (1991). *Continuous system modeling*. New York: Springer.
78. Chaldakov, G. N., Fiore, M., Ghenev, P. I., Beltowski, J., Rančić, G., Tunçel, N., & Aloe, L. (2014). Triactome: Neuro-immune-adipose interactions. Implication in vascular biology. *Frontiers in Immunology, 5*, 130.
79. Chalmers, D. J. (2006). Strong and weak emergence. In *The reemergence of emergence* (pp. 244–256). Oxford: Oxford University Press.
80. Chaos, A., Aldana, M., Espinosa-Soto, C., de León, B. G. P., Arroyo, A. G., & Alvarez-Buylla, E. R. (2006). From genes to flower patterns and evolution: Dynamic models of gene regulatory networks. *Journal of Plant Growth Regulation, 25*(4), 278–289.
81. Chaves, M., & Gouzé, J.-L. (2011). Exact control of genetic networks in a qualitative framework: The bistable switch example. *Automatica, 47*(6), 1105–1112.
82. Chaves, M., & Tournier, L. (2011). Predicting the asymptotic dynamics of large biological networks by interconnections of Boolean modules. In *2011 50th IEEE Conference on Decision and Control and European Control Conference (CDC-ECC)* (pp. 3026–3031). New York: IEEE.
83. Chen, C., & Wang, J. (2016). A physical mechanism of cancer heterogeneity. *Scientific Reports, 6*, 20679.
84. Chen, L., Liu, R., Liu, Z. P., Li, M., & Aihara, K. (2012). Detecting early-warning signals for sudden deterioration of complex diseases by dynamical network biomarkers. *Scientific Reports, 2*, 342.
85. Cheng, D., Qi, H., & Li, Z. (2010). *Analysis and control of Boolean networks: A semi-tensor product approach*. Berlin: Springer Science & Business Media.
86. Chickarmane, V., Troein, C., Nuber, U. A., Sauro, H. M., & Peterson, C. (2006). Transcriptional dynamics of the embryonic stem cell switch. *PLoS Computational Biology, 2*(9), e123.
87. Choi, M., Shi, J., Jung, S. H., Chen, X., & Cho, K. H. (2012). Attractor landscape analysis reveals feedback loops in the p53 network that control the cellular response to DNA damage. *Science Signaling, 5*(251), ra83.
88. Christodoulides, P., Hirata, Y., Domínguez-Hüttinger, E., Danby, S. G., Cork, M. J., Williams, H. C., et al. (2017). Computational design of treatment strategies for proactive therapy on atopic dermatitis using optimal control theory. *Philosophical Transactions of the Royal Society A, 375*(2096), 20160285.
89. Ciofani, M., Madar, A., Galan, C., Sellars, M., Mace, K., Pauli, F., et al. (2012). A validated regulatory network for Th17 cell specification. *Cell, 151*(2), 289–303.
90. Clark, W. H. (1995). The nature of cancer: Morphogenesis and progressive (self-)disorganization in neoplastic development and progression. *Acta Oncologica, 34*(1), 3–21.
91. Clausznitzer, D., Oleksiuk, O., Løvdok, L., Sourjik, V., & Endres, R. G. (2010). Chemotactic response and adaptation dynamics in Escherichia coli. *PLoS Computational Biology, 6*(5), e1000784.
92. Clune, J., Mouret, J. B., & Lipson, H. (2013). The evolutionary origins of modularity. *Proceedings of the Royal Society B, 280*(1755), 20122863.
93. Colman-Lerner, A., Gordon, A., Serra, E., Chin, T., Resnekov, O., Endy, D., et al. (2005). Regulated cell-to-cell variation in a cell-fate decision system. *Nature, 437*(7059), 699–706.
94. Conley, S. J., Gheordunescu, E., Kakarala, P., Newman, B., Korkaya, H., Heath, A. N., et al. (2012). Antiangiogenic agents increase breast cancer stem cells via the generation of tumor hypoxia. *Proceedings of the National Academy of Sciences of the United States of America, 109*(8), 2784–2789.

95. Cookson, W. (2004). The immunogenetics of asthma and eczema: A new focus on the epithelium. *Nature Reviews Immunology, 4*(12), 978–988.

96. Coppé, J. P., Patil, C. K., Rodier, F., Sun, Y., Muñoz, D. P., Goldstein, J., et al. (2008). Senescence-associated secretory phenotypes reveal cell-nonautonomous functions of onco-genic RAS and the p53 tumor suppressor. *PLoS Biology, 6*(12), e301.

97. Corblin, F., Fanchon, E., & Trilling, L. (2010). Applications of a formal approach to decipher discrete genetic networks. *BMC Bioinformatics, 11*(1), 385.

98. Cork, M. J., Danby, S. G., Vasilopoulos, Y., Hadgraft, J., Lane, M. E., Moustafa, M., et al. (2009). Epidermal barrier dysfunction in atopic dermatitis. *Journal of Investigative Dermatology, 129*(8), 1892–1908.

99. Cork, M. J., Robinson, D. A., Vasilopoulos, Y., Ferguson, A., Moustafa, M., MacGowan, A., et al. (2006). New perspectives on epidermal barrier dysfunction in atopic dermatitis: Gene–environment interactions. *Journal of Allergy and Clinical Immunology, 118*(1), 3–21.

100. Cosentino, C., & Bates, D. (2011). *Feedback control in systems biology.* Boca Raton: CRC Press.

101. Craciun, G., & Feinberg, M. (2005). Multiple equilibria in complex chemical reaction networks: I. The injectivity property. *SIAM Journal on Applied Mathematics, 65*(5), 1526–1546.

102. Creixell, P., Schoof, E. M., Erler, J. T., & Linding, R. (2012). Navigating cancer network attractors for tumor-specific therapy. *Nature Biotechnology, 30*(9), 842–848.

103. Crotty, S. (2011). Follicular helper CD4 T cells (Tfh). *Annual Review of Immunology, 29*, 621–663.

104. Cruickshanks, H. A., McBryan, T., Nelson, D. M., VanderKraats, N. D., Shah, P. P., Van Tuyn, J., et al. (2013). Senescent cells harbour features of the cancer epigenome. *Nature Cell Biology, 15*(12), 1495–1506.

105. Csermely, P., & Korcsmáros, T. (2013). Cancer-related networks: A help to understand, predict and change malignant transformation. *Seminars in Cancer Biology, 23*(4), 209–212.

106. Dai, L., Korolev, K. S., & Gore, J. (2013). Slower recovery in space before collapse of connected populations. *Nature, 496*(7445), 355–358.

107. Danby, S. G., Chittock, J., Brown, K., Albenali, L. H., & Cork, M. J. (2014). The effect of tacrolimus compared with betamethasone valerate on the skin barrier in volunteers with quiescent atopic dermatitis. *British Journal of Dermatology, 170*(4), 914–921.

108. Das, J., Chen, C. H., Yang, L., Cohn, L., Ray, P., & Ray, A. (2001). A critical role for NF-$\kappa$B in GATA3 expression and TH2 differentiation in allergic airway inflammation. *Nature Immunology, 2*(1), 45–50.

109. Davila-Velderrain, J., & Álvarez-Buylla, E. R. (2014). Bridging genotype and phenotype. In *Frontiers in ecology, evolution and complexity* (pp. 144–154). Mexico: EditoraC3 CopIt-arXives, UNAM.

110. Davila-Velderrain, J., Juarez-Ramiro, L., Martinez-Garcia, J. C., Alvarez-Buylla, E. R. (2015). Methods for characterizing the epigenetic attractors landscape associated with Boolean gene regulatory networks. arXiv preprint. arXiv:1510.04230.

111. Davila-Velderrain, J., Martinez-Garcia, J. C., & Álvarez-Buylla, E. R. (2015). Modeling the epigenetic attractors landscape: Toward a post-genomic mechanistic understanding of development. *Frontiers in Genetics, 6*, 160.

112. Davila-Velderrain, J., Villarreal, C., & Álvarez-Buylla, E. R. (2015). Reshaping the epigenetic landscape during early flower development: Induction of attractor transitions by relative differences in gene decay rates. *BMC Systems Biology, 9*(1), 20.

113. De Benedetto, A., Kubo, A., & Beck, L. A. (2012). Skin barrier disruption: A requirement for allergen sensitization? *Journal of Investigative Dermatology, 132*(3), 949–963.

114. De Craene, B., & Berx, G. (2013). Regulatory networks defining EMT during cancer initiation and progression. *Nature Reviews Cancer, 13*(2), 97–110.

115. De Jong, H., Geiselmann, J., Hernandez, C., & Page, M. (2003). Genetic network analyzer: Qualitative simulation of genetic regulatory networks. *Bioinformatics, 19*(3), 336–344.

116. de Luis Balaguer, M. A., Fisher, A. P., Clark, N. M., Fernandez-Espinosa, M. G., Moller, B., Weijers, D., et al. (2017). An inference approach combines spatial and temporal gene expression data to predict gene regulatory networks in Arabidopsis stem cells. bioRxiv, 140269.

117. Demerjian, M., Hachem, J. P., Tschachler, E., Denecker, G., Declercq, W., Vandenabeele, P., et al. (2008). Acute modulations in permeability barrier function regulate epidermal cornification: Role of caspase-14 and the protease-activated receptor type 2. *The American Journal of Pathology, 172*(1), 86–97.

118. Deraison, C., Bonnart, C., Lopez, F., Besson, C., Robinson, R., Jayakumar, A., et al. (2007). LEKTI fragments specifically inhibit KLK5, KLK7, and KLK14 and control desquamation through a pH-dependent interaction. *Molecular Biology of the Cell, 18*(9), 3607–3619.

119. Dolinoy, D. C., Weidman, J. R., & Jirtle, R. L. (2007). Epigenetic gene regulation: Linking early developmental environment to adult disease. *Reproductive Toxicology, 23*(3), 297–307.

120. Domínguez Hüttinger, E. (2014). Mathematical modelling of epithelium homeostasis (Doctoral dissertation, Imperial College London).

121. Dominguez-Hüttinger, E., Boon, N. J., Clarke, T. B., & Tanaka, R. J. (2017). Mathematical modelling of colonization, invasive infection and treatment of Streptococcus pneumoniae. *Frontiers in Physiology, 8*, 115.

122. Domínguez-Hüttinger, E., Christodoulides, P., Miyauchi, K., Irvine, A. D., Okada-Hatakeyama, M., Kubo, M., et al. (2017). Mathematical modeling of atopic dermatitis reveals "double-switch" mechanisms underlying 4 common disease phenotypes. *Journal of Allergy and Clinical Immunology, 139*(6), 1861–1872.

123. Domínguez-Hüttinger, E., Ono, M., Barahona, M., & Tanaka, R. J. (2013). Risk factor-dependent dynamics of atopic dermatitis: Modelling multi-scale regulation of epithelium homeostasis. *Interface Focus, 3*(2), 20120090.

124. Dommisch, H., Chung, W. O., Rohani, M. G., Williams, D., Rangarajan, M., Curtis, M. A., et al. (2007). Protease-activated receptor 2 mediates human beta-defensin 2 and CC chemokine ligand 20 mRNA expression in response to proteases secreted by Porphyromonas gingivalis. *Infection and Immunity, 75*(9), 4326–4333.

125. Doncic, A., Atay, O., Valk, E., Grande, A., Bush, A., Vasen, G., et al. (2015). Compartmentalization of a bistable switch enables memory to cross a feedback-driven transition. *Cell, 160*(6), 1182–1195.

126. Duhen, T., Duhen, R., Lanzavecchia, A., Sallusto, F., & Campbell, D. J. (2012). Functionally distinct subsets of human FOXP3+ Treg cells that phenotypically mirror effector Th cells. *Blood, 119*(19), 4430–4440.

127. DuPage, M., & Bluestone, J. A. (2016). Harnessing the plasticity of CD4+ T cells to treat immune-mediated disease. *Nature Reviews Immunology, 16*(3), 149–163.

128. Ehlken, B., Mhrenschlager, M., Kugland, B., Berger, K., Quednau, K., & Ring, J. (2005). Cost-of-illness study in patients suffering from atopic eczema in Germany. *Der Hautarzt; Zeitschrift für Dermatologie, Venerologie, und verwandte Gebiete, 56*(12), 1144–1151.

129. Eissing, T., Conzelmann, H., Gilles, E. D., Allgöwer, F., Bullinger, E., & Scheurich, P. (2004). Bistability analyses of a caspase activation model for receptor-induced apoptosis. *Journal of Biological Chemistry, 279*(35), 36892–36897.

130. Elias, P. M., Ahn, S. K., Denda, M., Brown, B. E., Crumrine, D., Kimutai, L. K., et al. (2002). Modulations in epidermal calcium regulate the expression of differentiation-specific markers. *Journal of Investigative Dermatology, 119*(5), 1128–1136.

131. Elias, P. M., Brown, B. E., Crumrine, D., Feingold, K. R., & Ahn, S. K. (2002). Origin of the epidermal calcium gradient: Regulation by barrier status and role of active vs passive mechanisms. *Journal of Investigative Dermatology, 119*(6), 1269–1274.

132. Elias, P. M., Hatano, Y., & Williams, M. L. (2008). Basis for the barrier abnormality in atopic dermatitis: Outside-inside-outside pathogenic mechanisms. *Journal of Allergy and Clinical Immunology, 121*(6), 1337–1343.

133. Elias, P. M., & Schmuth, M. (2009). Abnormal skin barrier in the etiopathogenesis of atopic dermatitis. *Current Allergy and Asthma Reports, 9*(4), 265–272.

134. Elias, P. M., & Wakefield, J. S. (2011). Therapeutic implications of a barrier-based pathogenesis of atopic dermatitis. *Clinical Reviews in Allergy & Immunology, 41*(3), 282–295.

135. Elinav, E., Nowarski, R., Thaiss, C. A., Hu, B., Jin, C., & Flavell, R. A. (2013). Inflammation-induced cancer: Crosstalk between tumours, immune cells and microorganisms. *Nature Reviews Cancer, 13*(11), 759–771.

136. Ellis, C. N., Drake, L. A., Prendergast, M. M., Abramovits, W., Boguniewicz, M., Daniel, C. R., et al. (2002). Cost of atopic dermatitis and eczema in the United States. *Journal of the American Academy of Dermatology, 46*(3), 361–370.

137. Ellmeier, W., & Taniuchi, I. (Eds.). (2014). *Transcriptional control of lineage differentiation in immune cells* (Vol. 381). Berlin: Springer.

138. Ellner, S. P., & Guckenheimer, J. (2011). *Dynamic models in biology*. Princeton: Princeton University Press.

139. Elnashaie, S. S., & Uhlig, F. (2007). *Numerical techniques for chemical and biological engineers using MATLAB: A simple bifurcation approach*. Berlin: Springer Science & Business Media.

140. Enciso, J., Álvarez-Buylla, E., & Pelayo, R. (2017). A multi-modular Boolean network for the study of acute lymphoblastic leukemia. *Experimental Hematology, 53*, S109.

141. Espinosa-Soto, C., Padilla-Longoria, P., & Álvarez-Buylla, E. R. (2004). A gene regulatory network model for cell–fate determination during Arabidopsis thaliana flower development that is robust and recovers experimental gene expression profiles. *Plant Cell, 16*, 2923–2939.

142. Ewald, J. A., Desotelle, J. A., Wilding, G., & Jarrard, D. F. (2010). Therapy-induced senescence in cancer. *Journal of the National Cancer Institute, 102*(20), 1536–1546.

143. Fallon, P. G., Sasaki, T., Sandilands, A., Campbell, L. E., Saunders, S. P., Mangan, N. E., et al. (2009). A homozygous frameshift mutation in the mouse Flg gene facilitates enhanced percutaneous allergen priming. *Nature Genetics, 41*(5), 602–608.

144. Fauré, A., Naldi, A., Chaouiya, C., & Thieffry, D. (2006). Dynamical analysis of a generic Boolean model for the control of the mammalian cell cycle. *Bioinformatics, 22*(14), e124–e131.

145. Feinberg, A. P., Koldobskiy, M. A., & Göndör, A. (2016). Epigenetic modulators, modifiers and mediators in cancer aetiology and progression. *Nature Reviews Genetics, 17*(5), 284–299.

146. Ferrell, J. E. (2012). Bistability, bifurcations, and Waddington's epigenetic landscape. *Current Biology, 22*(11), R458–R466.

147. Ferrell, J. E., & Machleder, E. M. (1998). The biochemical basis of an all-or-none cell fate switch in Xenopus oocytes. *Science, 280*(5365), 895–898.

148. Ferrell, J. E., Pomerening, J. R., Kim, S. Y., Trunnell, N. B., Xiong, W., Huang, C. Y. F., et al. (2009). Simple, realistic models of complex biological processes: Positive feedback and bistability in a cell fate switch and a cell cycle oscillator. *FEBS Letters, 583*(24), 3999–4005.

149. Fey, D., Halasz, M., Dreidax, D., Kennedy, S. P., Hastings, J. F., Rauch, N., et al. (2015). Signaling pathway models as biomarkers: Patient-specific simulations of JNK activity predict the survival of neuroblastoma patients. *Science Signaling, 8*(408), 1–16.

150. Flintoft, L. (2005). From genotype to phenotype: A shortcut through the library. *Nature Reviews Genetics, 6*. Article ID 520.

151. Foo, J., Leder, K., & Michor, F. (2011). Stochastic dynamics of cancer initiation. *Physical Biology, 8*(1), 015002.

152. Fortugno, P., Furio, L., Teson, M., Berretti, M., El Hachem, M., Zambruno, G., et al. (2012). The 420K LEKTI variant alters LEKTI proteolytic activation and results in protease deregulation: Implications for atopic dermatitis. *Human Molecular Genetics, 21*(19), 4187–4200.

153. Fox, R. F. (1993). Review of Stuart Kauffman, The origins of order: Self-organization and selection in evolution. *Biophysical Journal, 65*(6), 2698.
154. Franks, P. W., Pearson, E., & Florez, J. C. (2013). Gene-environment and gene-treatment interactions in type 2 diabetes. *Diabetes Care, 36*(5), 1413–1421.
155. Friend, S. H., Dryja, T. P., & Weinberg, R. A. (1988). Oncogenes and tumor-suppressing genes. *New England Journal of Medicine, 318*(10), 618–622.
156. Fry, L. (2003). *An atlas of atopic eczema.* Boca Raton: Parthenon.
157. Fuchs, A. (2014). *Nonlinear dynamics in complex systems.* Berlin: Springer.
158. Galic, S., Sachithanandan, N., Kay, T. W., & Steinberg, G. R. (2014). Suppressor of cytokine signalling (SOCS) proteins as guardians of inflammatory responses critical for regulating insulin sensitivity. *Biochemical Journal, 461*(2), 177–188.
159. Gallagher, K. L., Sozzani, R., & Lee, C. M. (2014). Intercellular protein movement: Deciphering the language of development. *Annual Review of Cell and Developmental Biology, 30*, 207–233.
160. Galli, S. J., Tsai, M., & Piliponsky, A. M. (2008). The development of allergic inflammation. *Nature, 454*(7203), 445–454.
161. Gao, J., Barzel, B., & Barabási, A. L. (2016). Universal resilience patterns in complex networks. *Nature, 530*(7590), 307–312.
162. Garbe, J. C., Vrba, L., Sputova, K., Fuchs, L., Novak, P., Brothman, A. R., et al. (2014). Immortalization of normal human mammary epithelial cells in two steps by direct targeting of senescence barriers does not require gross genomic alterations. *Cell Cycle, 13*(21), 3423–3435.
163. Garg, A., Mohanram, K., De Micheli, G., & Xenarios, I. (2012). Implicit methods for qualitative modeling of gene regulatory networks. In *Gene regulatory networks: Methods and protocols* (Vol. 786, pp. 397–443). New York: Humana Press.
164. Garraway, L. A., & Lander, E. S. (2013). Lessons from the cancer genome. *Cell, 153*(1), 17–37.
165. Geginat, J., Paroni, M., Facciotti, F., Gruarin, P., Kastirr, I., Caprioli, F., et al. (2013). The CD4-centered universe of human T cell subsets. *Seminars in Immunology, 25*(4), 252–262. New York: Academic.
166. Gene Ontology Consortium. (2004). The Gene Ontology (GO) database and informatics resource. *Nucleic Acids Research, 32*(Suppl. 1), D258–D261.
167. Gershenfeld, N. (1998). *The nature of mathematical modeling.* Cambridge: Cambridge University Press.
168. Gittler, J. K., Shemer, A., Suárez-Fariñas, M., Fuentes-Duculan, J., Gulewicz, K. J., Wang, C. Q., et al. (2012). Progressive activation of T H 2/T H 22 cytokines and selective epidermal proteins characterizes acute and chronic atopic dermatitis. *Journal of Allergy and Clinical Immunology, 130*(6), 1344–1354.
169. Glass, L., & Kauffman, S. A. (1973). The logical analysis of continuous, non-linear biochemical control networks. *Journal of Theoretical Biology, 39*(1), 103–129.
170. Goldberg, M. (2008). A systematic review of the relation between long-term exposure to ambient air pollution and chronic diseases. *Reviews on Environmental Health, 23*(4), 243–298.
171. Goldbeter, A., & Koshland, D. E. (1984). Ultrasensitivity in biochemical systems controlled by covalent modification. Interplay between zero-order and multistep effects. *Journal of Biological Chemistry, 259*(23), 14441–14447.
172. Goodwin, B. C. (1994). How the leopard changed its spots. In *The evolution of complexity.* Princeton: Princeton University Press.
173. Goodwin, B. C. (2009). Beyond the Darwinian paradigm: Understanding biological forms. In *Evolution: The first four billion years* (pp. 299–312). Cambridge: Harvard University Press.
174. Gordillo, M., Evans, T., & Gouon-Evans, V. (2015). Orchestrating liver development. *Development, 142*(12), 2094–2108.
175. Goswami, R., & Kaplan, M. H. (2011). A brief history of IL-9. *The Journal of Immunology, 186*(6), 3283–3288.

176. Gould, H. J., & Sutton, B. J. (2008). IgE in allergy and asthma today. *Nature Reviews Immunology, 8*(3), 205–217.
177. Grabe, N., & Neuber, K. (2005). A multicellular systems biology model predicts epidermal morphology, kinetics and Ca2+ flow. *Bioinformatics, 21*(17), 3541–3547.
178. Grarup, N., Sandholt, C. H., Hansen, T., & Pedersen, O. (2014). Genetic susceptibility to type 2 diabetes and obesity: From genome-wide association studies to rare variants and beyond. *Diabetologia, 57*(8), 1528–1541.
179. Grazia Roncarolo, M., Gregori, S., Battaglia, M., Bacchetta, R., Fleischhauer, K., & Levings, M. K. (2006). Interleukin-10-secreting type 1 regulatory T cells in rodents and humans. *Immunological Reviews, 212*(1), 28–50.
180. Green, E. D., Rubin, E. M., & Olson, M. V. (2017). The future of DNA sequencing. *Nature News, 550*(7675), 179.
181. Groß-Hardt, R., & Laux, T. (2003). Stem cell regulation in the shoot meristem. *Journal of Cell Science, 116*(9), 1659–1666.
182. Grosso, G., Bella, F., Godos, J., Sciacca, S., Del Rio, D., Ray, S., et al. (2017). Possible role of diet in cancer: Systematic review and multiple meta-analyses of dietary patterns, lifestyle factors, and cancer risk. *Nutrition Reviews, 75*(6), 405–419.
183. Guevara, M. R. (2003). Bifurcations involving fixed points and limit cycles in biological systems. In *Dynamics in physiology and medicine* (pp. 41–85). New York: Springer.
184. Hachem, J. P., Crumrine, D., Fluhr, J., Brown, B. E., Feingold, K. R., & Elias, P. M. (2003). pH directly regulates epidermal permeability barrier homeostasis, and stratum corneum integrity/cohesion. *Journal of Investigative Dermatology, 121*(2), 345–353.
185. Hachem, J. P., Houben, E., Crumrine, D., Man, M. Q., Schurer, N., Roelandt, T., et al. (2006). Serine protease signaling of epidermal permeability barrier homeostasis. *Journal of Investigative Dermatology, 126*(9), 2074–2086.
186. Hahn, W. C., & Weinberg, R. A. (2002). Rules for making human tumor cells. *New England Journal of Medicine, 347*(20), 1593–1603.
187. Halasz, M., Kholodenko, B. N., Kolch, W., & Santra, T. (2016). Integrating network reconstruction with mechanistic modeling to predict cancer therapies. *Science Signaling, 9*(455), ra114.
188. Ham, H. J., Andeweg, A. C., & Boer, R. J. (2013). Induction of appropriate Th-cell phenotypes: Cellular decision-making in heterogeneous environments. *Parasite Immunology, 35*(11), 318–330.
189. Hamant, O., Traas, J., & Boudaoud, A. (2010). Regulation of shape and patterning in plant development. *Current Opinion in Genetics & Development, 20*(4), 454–459.
190. Hammad, H., & Lambrecht, B. N. (2008). Dendritic cells and epithelial cells: Linking innate and adaptive immunity in asthma. *Nature Reviews Immunology, 8*(3), 193–204.
191. Han, J. M., Patterson, S. J., Speck, M., Ehses, J. A., & Levings, M. K. (2014). Insulin inhibits IL-10-mediated regulatory T cell function: Implications for obesity. *The Journal of Immunology, 192*(2), 623–629.
192. Hanahan, D., & Weinberg, R. A. (2011). Hallmarks of cancer: The next generation. *Cell, 144*(5), 646–674.
193. Hansemann, D. (1890). Ueber asymmetrische Zelltheilung in Epithelkrebsen und deren biologische Bedeutung. *Virchows Archiv, 119*(2), 299–326.
194. Harrington, H. A., Ho, K. L., Ghosh, S., & Tung, K. C. (2008). Construction and analysis of a modular model of caspase activation in apoptosis. *Theoretical Biology and Medical Modelling, 5*(1), 26.
195. Hijmans, B. S., Tiemann, C. A., Grefhorst, A., Boesjes, M., van Dijk, T. H., Tietge, U. J., et al. (2015). A systems biology approach reveals the physiological origin of hepatic steatosis induced by liver X receptor activation. *The FASEB Journal, 29*(4), 1153–1164.
196. Hinrichsen, D., & Pritchard, A. J. (2005). *Mathematical Systems Theory I - Modelling, State Space Analysis, Stability and Robustness. Texts in applied mathematics* (Vol. 48). Berlin: Springer.

197. Hinz, B. (2009). Tissue stiffness, latent TGF-$\beta$1 activation, and mechanical signal transduction: Implications for the pathogenesis and treatment of fibrosis. *Current Rheumatology Reports, 11*(2), 120–126.

198. Hirata, Y., Bruchovsky, N., & Aihara, K. (2010). Development of a mathematical model that predicts the outcome of hormone therapy for prostate cancer. *Journal of Theoretical Biology, 264*(2), 517–527.

199. Hirata, Y., di Bernardo, M., Bruchovsky, N., & Aihara, K. (2010). Hybrid optimal scheduling for intermittent androgen suppression of prostate cancer. *Chaos: An Interdisciplinary Journal of Nonlinear Science, 20*(4), 045125.

200. Hirata, Y., Morino, K., Akakura, K., Higano, C. S., Bruchovsky, N., Gambol, T., et al. (2015). Intermittent androgen suppression: Estimating parameters for individual patients based on initial PSA data in response to androgen deprivation therapy. *PloS One, 10*(6), e0130372.

201. Höfer, T., Nathansen, H., Löhning, M., Radbruch, A., & Heinrich, R. (2002). GATA-3 transcriptional imprinting in Th2 lymphocytes: A mathematical model. *Proceedings of the National Academy of Sciences, 99*(14), 9364–9368.

202. Hoffman, D. R., Kroll, L. M., Basehoar, A., Reece, B., Cunningham, C. T., & Koenig, D. W. (2014). Immediate and extended effects of sodium lauryl sulphate exposure on stratum corneum natural moisturizing factor. *International Journal of Cosmetic Science, 36*(1), 93–101.

203. Hoffman, D. R., Kroll, L. M., Basehoar, A., Reece, B., Cunningham, C. T., & Koenig, D. W. (2015). Immediate and extended effects of abrasion on stratum corneum natural moisturizing factor. *Skin Research and Technology, 21*(3), 366–372.

204. Hollstein, M., Sidransky, D., Vogelstein, B., & Harris, C. C. (1991). p53 mutations in human cancers. *Science, 253*(5015), 49–53.

205. Hong, T., Watanabe, K., Ta, C. H., Villarreal-Ponce, A., Nie, Q., & Dai, X. (2015). An Ovol2-Zeb1 mutual inhibitory circuit governs bidirectional and multi-step transition between epithelial and mesenchymal states. *PLoS Computational Biology, 11*(11), e1004569.

206. Horimukai, K., Morita, K., Narita, M., Kondo, M., Kitazawa, H., Nozaki, M., et al. (2014). Application of moisturizer to neonates prevents development of atopic dermatitis. *Journal of Allergy and Clinical Immunology, 134*(4), 824–830.

207. Horn, H., Lawrence, M. S., Chouinard, C. R., Shrestha, Y., Hu, J. X., Worstell, E., et al. (2017). Expanding discovery from cancer genomes by integrating protein network analyses with in vivo tumorigenesis assays. bioRxiv, 151977.

208. Hovnanian, A. (2013). Netherton syndrome: Skin inflammation and allergy by loss of protease inhibition. *Cell and Tissue Research, 351*(2), 289–300.

209. Howell, M. D., Kim, B. E., Gao, P., Grant, A. V., Boguniewicz, M., DeBenedetto, A., et al. (2007). Cytokine modulation of atopic dermatitis filaggrin skin expression. *Journal of Allergy and Clinical Immunology, 120*(1), 150–155.

210. Howes, A., Stimpson, P., Redford, P., Gabrysova, L., & O'Garra, A. (2014). Interleukin-10: Cytokines in anti-inflammation and tolerance. In *Cytokine frontiers* (pp. 327–352). Tokyo: Springer.

211. Huang, S. (2011). Systems biology of stem cells: Three useful perspectives to help overcome the paradigm of linear pathways. *Philosophical Transactions of the Royal Society of London B: Biological Sciences, 366*(1575), 2247–2259.

212. Huang, S. (2011). On the intrinsic inevitability of cancer: From foetal to fatal attraction. *Seminars in Cancer Biology, 21*(3), 183–199.

213. Huang, S. (2013). Genetic and non-genetic instability in tumor progression: Link between the fitness landscape and the epigenetic landscape of cancer cells. *Cancer and Metastasis Reviews, 32*(3–4), 423–448.

214. Huang, S., Eichler, G., Bar-Yam, Y., & Ingber, D. E. (2005). Cell fates as high-dimensional attractor states of a complex gene regulatory network. *Physical Review Letters, 94*(12), 128701.

215. Huang, S., Ernberg, I., & Kauffman, S. (2009). Cancer attractors: A systems view of tumors from a gene network dynamics and developmental perspective. *Seminars in Cell & Developmental Biology, 20*(7), 869–876.
216. Huang, S., & Ingber, D. E. (2007). A non-genetic basis for cancer progression and metastasis: Self-organizing attractors in cell regulatory networks. *Breast Disease, 26*(1), 27–54.
217. Huang, S., & Kauffman, S. A. (2009). Complex GRN regulatory networks: From structure to biological observables: Cell fate determination. In *Encyclopedia of complexity and systems science* (pp. 1180–1213). New York: Springer.
218. Hucka, M., Finney, A., Sauro, H. M., Bolouri, H., Doyle, J. C., Kitano, H., et al. (2003). The systems biology markup language (SBML). A medium for representation and exchange of biochemical network models. *Bioinformatics, 19*(4), 524–531.
219. Hudson, T. J., Anderson, W., Aretz, A., Barker, A. D., Bell, C., Bernabé, R. R., et al. (2010). International network of cancer genome projects. *Nature, 464*(7291), 993–998.
220. Ideta, A. M., Tanaka, G., Takeuchi, T., & Aihara, K. (2008). A mathematical model of intermittent androgen suppression for prostate cancer. *Journal of Nonlinear Science, 18*(6), 593–614.
221. Inoue, K., Shinohara, H., Behar, M., Yumoto, N., Tanaka, G., Hoffmann, A., et al. (2016). Oscillation dynamics underlie functional switching of NF-$\kappa$B for B-cell activation. *NPJ Systems Biology and Applications, 2*, 16024.
222. Irvine, A. D., McLean, W. I., & Leung, D. Y. (2011). Filaggrin mutations associated with skin and allergic diseases. *New England Journal of Medicine, 365*(14), 1315–1327.
223. Ivanova, E. A., & Orekhov, A. N. (2015). T helper lymphocyte subsets and plasticity in autoimmunity and cancer: An overview. *BioMed Research International, 2015*, 9. Article ID 327470.
224. Izhikevich, E. M. (2007). *Dynamical systems in neuroscience*. Cambridge: MIT Press.
225. Jeanquartier, F., Jean-Quartier, C., Kotlyar, M., Tokar, T., Hauschild, A. C., Jurisica, I., et al. (2016). Machine learning for in silico modeling of tumor growth. In *Machine learning for health informatics* (pp. 415–434). Berlin: Springer International Publishing.
226. Jensen, K. J., Moyer, C. B., & Janes, K. A. (2016). Network architecture predisposes an enzyme to either pharmacologic or genetic targeting. *Cell Systems, 2*(2), 112–121.
227. Jeong, H., Mason, S. P., Barabási, A. L., & Oltvai, Z. N. (2001). Lethality and centrality in protein networks. *Nature, 411*(6833), 41–42.
228. Juárez-Ramiro, L. (2015). *Modelado de patrones de interconexión de módulos de regulación transcripcional: aplicación a la interacción entre la transición epitelio-mesénquima y el ciclo celular.* Tesis de Maestría en Ciencias, Agosto, Departamento de Control Automático, Centro de Investigación y de Estudios Avanzados del Instituto Politécnico Nacional.
229. Jurtz, V. I., Johansen, A. R., Nielsen, M., Almagro Armenteros, J. J., Nielsen, H., Sønderby, C. K., et al. (2017). An introduction to deep learning on biological sequence data: Examples and solutions. *Bioinformatics, 33*(22), 3685–3690.
230. Kalluri, R. (2009). EMT: When epithelial cells decide to become mesenchymal-like cells. *The Journal of Clinical Investigation, 119*(6), 1417.
231. Kanehisa, M., Furumichi, M., Tanabe, M., Sato, Y., & Morishima, K. (2017). KEGG: New perspectives on genomes, pathways, diseases and drugs. *Nucleic Acids Research, 45*(D1), D353–D361.
232. Kaneko, K. (2006). *Life: An introduction to complex systems biology*. Berlin: Springer.
233. Kaneko, K. (2011). Characterization of stem cells and cancer cells on the basis of gene expression profile stability, plasticity, and robustness. *Bioessays, 33*(6), 403–413.
234. Kanno, Y., Vahedi, G., Hirahara, K., Singleton, K., & O'Shea, J. J. (2012). Transcriptional and epigenetic control of T helper cell specification: Molecular mechanisms underlying commitment and plasticity. *Annual Review of Immunology, 30*, 707–731.
235. Kaplan, D., & Glass, L. (2012). *Understanding nonlinear dynamics*. Berlin: Springer Science & Business Media.
236. Kaplan, D. H., Igyártó, B. Z., & Gaspari, A. A. (2012). Early immune events in the induction of allergic contact dermatitis. *Nature Reviews Immunology, 12*(2), 114–124.

237. Kaplan, M. H. (2013). Th9 cells: Differentiation and disease. *Immunological Reviews, 252*(1), 104–115.
238. Kauffman, S. (1971). Gene regulation networks: A theory for their global structure and behaviors. *Current Topics in Developmental Biology, 6,* 145–182.
239. Kauffman, S. A. (1969). Homeostasis and differentiation in random genetic control networks. *Nature, 224*(5215), 177–178.
240. Kauffman, S. A. (1969). Metabolic stability and epigenesis in randomly constructed genetic nets. *Journal of Theoretical Biology, 22*(3), 437–467.
241. Kauffman, S. A. (1993). *The origins of order: Self-organization and selection in evolution.* Oxford: Oxford University Press.
242. Kawasaki, H., Nagao, K., Kubo, A., Hata, T., Shimizu, A., Mizuno, H., et al. (2012). Altered stratum corneum barrier and enhanced percutaneous immune responses in filaggrin-null mice. *Journal of Allergy and Clinical Immunology, 129*(6), 1538–1546.
243. Keller, A. D. (1995). Model genetic circuits encoding autoregulatory transcription factors. *Journal of Theoretical Biology, 172*(2), 169–185.
244. Kennedy, M. A., Gonzalez-Sarmiento, R., Kees, U. R., Lampert, F., Dear, N., Boehm, T., et al. (1991). HOX11, a homeobox-containing T-cell oncogene on human chromosome 10q24. *Proceedings of the National Academy of Sciences, 88*(20), 8900–8904.
245. Kezic, S., O'regan, G. M., Yau, N., Sandilands, A., Chen, H., Campbell, L. E., et al. (2011). Levels of filaggrin degradation products are influenced by both filaggrin genotype and atopic dermatitis severity. *Allergy, 66*(7), 934–940.
246. Kim, Y. A., & Przytycka, T. M. (2013). Bridging the gap between genotype and phenotype via network approaches. *Frontiers in Genetics, 3,* 227.
247. Kirkpatrick, S., Gelatt, C. D., Jr., & Vecchi, M. P. (1987). Optimization by simulated annealing. In *Spin glass theory and beyond: An introduction to the replica method and its applications, World scientific lecture notes in physics* (Vol. 9, pp. 339–348). Singapore: World Scientific.
248. Kleinewietfeld, M., & Hafler, D. A. (2013). The plasticity of human Treg and Th17 cells and its role in autoimmunity. *Seminars in Immunology, 25*(4), 305–312.
249. Knosp, C. A., & Johnston, J. A. (2012). Regulation of CD4+ T-cell polarization by suppressor of cytokine signalling proteins. *Immunology, 135*(2), 101–111.
250. Komatsu, N., Saijoh, K., Kuk, C., Liu, A. C., Khan, S., Shirasaki, F., et al. (2007). Human tissue kallikrein expression in the stratum corneum and serum of atopic dermatitis patients. *Experimental Dermatology, 16*(6), 513–519.
251. Kong, H. H., Oh, J., Deming, C., Conlan, S., Grice, E. A., Beatson, M. A., et al. (2012). Temporal shifts in the skin microbiome associated with disease flares and treatment in children with atopic dermatitis. *Genome Research, 22*(5), 850–859.
252. Krogan, N. J., Lippman, S., Agard, D. A., Ashworth, A., & Ideker, T. (2015). The cancer cell map initiative: Defining the hallmark networks of cancer. *Molecular Cell, 58*(4), 690–698.
253. Kundaje, A., Meuleman, W., Ernst, J., Bilenky, M., Yen, A., Heravi-Moussavi, A., et al. (2015). Integrative analysis of 111 reference human epigenomes. *Nature, 518*(7539), 317–330.
254. Kundu, J. K., & Surh, Y. J. (2008). Inflammation: Gearing the journey to cancer. *Mutation Research/Reviews in Mutation Research, 659*(1), 15–30.
255. Kuo, I. H., Yoshida, T., De Benedetto, A., & Beck, L. A. (2013). The cutaneous innate immune response in patients with atopic dermatitis. *Journal of Allergy and Clinical Immunology, 131*(2), 266–278.
256. Kuznetsov, Y. A. (2013). *Elements of applied bifurcation theory* (Vol. 112). Berlin: Springer Science & Business Media.
257. Lambrecht, B. N., & Hammad, H. (2010). The role of dendritic and epithelial cells as master regulators of allergic airway inflammation. *The Lancet, 376*(9743), 835–843.
258. Lanczos, C. (2012). *The variational principles of mechanics.* North Chelmsford: Courier Corporation.

259. Land, H., Parada, L. F., & Weinberg, R. A. (1983). Tumorigenic conversion of primary embryo fibroblasts requires at least two cooperating oncogenes. *Nature, 304*(5927), 596–602.

260. Lasry, A., & Ben-Neriah, Y. (2015). Senescence-associated inflammatory responses: Aging and cancer perspectives. *Trends in Immunology, 36*(4), 217–228.

261. Latham, A. M., Molina-París, C., Homer-Vanniasinkam, S., & Ponnambalam, S. (2010). An integrative model for vascular endothelial growth factor A as a tumour biomarker. *Integrative Biology, 2*(9), 397–407.

262. Lathia, J. D., Mack, S. C., Mulkearns-Hubert, E. E., Valentim, C. L., & Rich, J. N. (2015). Cancer stem cells in glioblastoma. *Genes & Development, 29*(12), 1203–1217.

263. Laubenbacher, R., Hower, V., Jarrah, A., Torti, S. V., Shulaev, V., Mendes, P., et al. (2009). A systems biology view of cancer. *Biochimica et Biophysica Acta—Reviews on Cancer, 1796*(2), 129–139.

264. Laurent, M., & Kellershohn, N. (1999). Multistability: A major means of differentiation and evolution in biological systems. *Trends in Biochemical Sciences, 24*(11), 418–422.

265. Le Novere, N., Hucka, M., Mi, H., Moodie, S., Schreiber, F., Sorokin, A., et al. (2009). The systems biology graphical notation. *Nature Biotechnology, 27*(8), 735–741.

266. Lee, S. E., Kim, J. M., Jeong, S. K., Jeon, J. E., Yoon, H. J., Jeong, M. K., et al. (2010). Protease-activated receptor-2 mediates the expression of inflammatory cytokines, antimicrobial peptides, and matrix metalloproteinases in keratinocytes in response to Propionibacterium acnes. *Archives of Dermatological Research, 302*(10), 745–756.

267. Lee, T. I., & Young, R. A. (2013). Transcriptional regulation and its misregulation in disease. *Cell, 152*(6), 1237–1251.

268. Lee, Y. K., Mukasa, R., Hatton, R. D., & Weaver, C. T. (2009). Developmental plasticity of Th17 and Treg cells. *Current Opinion in Immunology, 21*(3), 274–280.

269. Lengauer, C., Kinzler, K. W., & Vogelstein, B. (1998). Genetic instabilities in human cancers. *Nature, 396*(6712), 643–649.

270. Leung, D. Y. (2000). Atopic dermatitis: New insights and opportunities for therapeutic intervention. *Journal of Allergy and Clinical Immunology, 105*(5), 860–876.

271. Lewontin, R. (1974). *The genetic basis of evolutionary change. Columbia biological series.* Columbia: Columbia University Press.

272. Li, C., & Wang, J. (2013). Quantifying cell fate decisions for differentiation and reprogramming of a human stem cell network: Landscape and biological paths. *PLoS Computational Biology, 9*(8), e1003165.

273. Li, C., & Wang, J. (2014). Quantifying the underlying landscape and paths of cancer. *Journal of the Royal Society Interface, 11*(100), 20140774.

274. Li, C. W., Xia, W., Huo, L., Lim, S. O., Wu, Y., Hsu, J. L., et al. (2012). Epithelial–mesenchymal transition induced by TNF-$\alpha$ requires NF-$\kappa$B-mediated transcriptional upregulation of Twist1. *Cancer Research, 72*(5), 1290–1300.

275. Li, E., Materna, S. C., & Davidson, E. H. (2013). New regulatory circuit controlling spatial and temporal gene expression in the sea urchin embryo oral ectoderm GRN. *Developmental Biology, 382*(1), 268–279.

276. Liberman, L. M., Sozzani, R., & Benfey, P. N. (2012). Integrative systems biology: An attempt to describe a simple weed. *Current Opinion in Plant Biology, 15*(2), 162–167.

277. Lillacci, G., & Khammash, M. (2010). Parameter estimation and model selection in computational biology. *PLoS Computational Biology, 6*(3), e1000696.

278. Lindberg, M. J., Popko-Scibor, A. E., Hansson, M. L., & Wallberg, A. E. (2010). SUMO modification regulates the transcriptional activity of MAML1. *The FASEB Journal, 24*(7), 2396–2404.

279. Lipniacki, T., Hat, B., Faeder, J. R., & Hlavacek, W. S. (2008). Stochastic effects and bistability in T cell receptor signaling. *Journal of Theoretical Biology, 254*(1), 110–122.

280. Liston, A., & Gray, D. H. (2014). Homeostatic control of regulatory T cell diversity. *Nature Reviews Immunology, 14*(3), 154–165.

281. Liu, E. T., & Lauffenburger, D. A. (Eds.). (2009). *Systems biomedicine: Concepts and perspectives.* New York: Academic.

282. Liu, L., You, Z., Yu, H., Zhou, L., Zhao, H., Yan, X., et al. (2017). Mechanotransduction-modulated fibrotic microniches reveal the contribution of angiogenesis in liver fibrosis. *Nature Materials, 16*, 1252. nmat5024.

283. Liu, Z., Li, Z., Mao, K., Zou, J., Wang, Y., Tao, Z., et al. (2009). Dec2 promotes Th2 cell differentiation by enhancing IL-2R signaling. *The Journal of Immunology, 183*(10), 6320–6329.

284. Lobo, D., & Levin, M. (2015). Inferring regulatory networks from experimental morphological phenotypes: A computational method reverse-engineers planarian regeneration. *PLoS Computational Biology, 11*(6), e1004295.

285. Loeb, L. A., & Harris, C. C. (2008). Advances in chemical carcinogenesis: A historical review and prospective. *Cancer Research, 68*(17), 6863–6872.

286. Loktionov, A. (2003). Common gene polymorphisms and nutrition: Emerging links with pathogenesis of multifactorial chronic diseases. *The Journal of Nutritional Biochemistry, 14*(8), 426–451.

287. Longo, V. D., & Fontana, L. (2010). Calorie restriction and cancer prevention: Metabolic and molecular mechanisms. *Trends in Pharmacological Sciences, 31*(2), 89–98.

288. Longo, V. D., & Mattson, M. P. (2014). Fasting: Molecular mechanisms and clinical applications. *Cell Metabolism, 19*(2), 181–192.

289. Louis, M., & Becskei, A. (2002). Binary and graded responses in gene networks. *Science STKE, 2002*(143), pe33.

290. Lu, Y., Hong, S., Li, H., Park, J., Hong, B., Wang, L., et al. (2012). Th9 cells promote antitumor immune responses in vivo. *The Journal of Clinical Investigation, 122*(11), 4160.

291. Lynch, J., & Smith, G. D. (2005). A life course approach to chronic disease epidemiology. *Annual Review of Public Health, 26*, 1–35.

292. MacIver, N. J., Michalek, R. D., & Rathmell, J. C. (2013). Metabolic regulation of T lymphocytes. *Annual Review of Immunology, 31*, 259–283.

293. Magi, S., Iwamoto, K., & Okada-Hatakeyama, M. (2017). Current status of mathematical modeling of cancer – From the viewpoint of cancer hallmarks. *Current Opinion in Systems Biology, 2*, 39–48.

294. Mahalingam, D., Kong, C. M., Lai, J., Tay, L. L., Yang, H., & Wang, X. (2012). Reversal of aberrant cancer methylome and transcriptome upon direct reprogramming of lung cancer cells. *Scientific Reports, 2*, 592.

295. Mammoto, T., & Ingber, D. E. (2010). Mechanical control of tissue and organ development. *Development, 137*(9), 1407–1420.

296. Man, M. Q., Hatano, Y., Lee, S. H., Man, M., Chang, S., Feingold, K. R., et al. (2008). Characterization of a hapten-induced, murine model with multiple features of atopic dermatitis: Structural, immunologic, and biochemical changes following single versus multiple oxazolone challenges. *Journal of Investigative Dermatology, 128*(1), 79–86.

297. Mancini, A. J., Kaulback, K., & Chamlin, S. L. (2008). The socioeconomic impact of atopic dermatitis in the United States: A systematic review. *Pediatric Dermatology, 25*(1), 1–6.

298. Mani, S. A., Guo, W., Liao, M. J., Eaton, E. N., Ayyanan, A., Zhou, A. Y., et al. (2008). The epithelial-mesenchymal transition generates cells with properties of stem cells. *Cell, 133*(4), 704–715.

299. Marchiando, A. M., Graham, W. V., & Turner, J. R. (2010). Epithelial barriers in homeostasis and disease. *Annual Review of Pathological Mechanical Disease, 5*, 119–144.

300. Margueron, R., & Reinberg, D. (2010). Chromatin structure and the inheritance of epigenetic information. *Nature Reviews Genetics, 11*(4), 285.

301. Marino, S., Hogue, I. B., Ray, C. J., & Kirschner, D. E. (2008). A methodology for performing global uncertainty and sensitivity analysis in systems biology. *Journal of Theoretical Biology, 254*(1), 178–196.

302. Marioni, J. C., & Arendt, D. (2017). How single-cell genomics is changing evolutionary and developmental biology. *Annual Review of Cell and Developmental Biology, 33*(1), 537–553.

303. Martinez-Sanchez, M. E., Mendoza, L., Villarreal, C., & Álvarez-Buylla, E. R. (2015). A minimal regulatory network of extrinsic and intrinsic factors recovers observed patterns of CD4+ T cell differentiation and plasticity. *PLoS Computational Biology, 11*(6), e1004324.
304. Mayburd, A. L. (2009). Expression variation: Its relevance to emergence of chronic disease and to therapy. *PLoS One, 4*(6), e5921.
305. Mayo, A. E., Setty, Y., Shavit, S., Zaslaver, A., & Alon, U. (2006). Plasticity of the cis-regulatory input function of a gene. *PLoS Biology, 4*(4), e45.
306. McAleer, M. A., & Irvine, A. D. (2013). The multifunctional role of filaggrin in allergic skin disease. *Journal of Allergy and Clinical Immunology, 131*(2), 280–291.
307. McDonnell, M. D., & Abbott, D. (2009). What is stochastic resonance?. Definitions, misconceptions, debates, and its relevance to biology. *PLoS Computational Biology, 5*(5), e1000348.
308. McGeer, P. L., & Mcgeer, E. G. (2004). Inflammation and the degenerative diseases of aging. *Annals of the New York Academy of Sciences, 1035*(1), 104–116.
309. McGirt, L. Y., & Beck, L. A. (2006). Innate immune defects in atopic dermatitis. *Journal of Allergy and Clinical Immunology, 118*(1), 202–208.
310. McLaughlin, T., Liu, L. F., Lamendola, C., Shen, L., Morton, J., Rivas, H., et al. (2014). T-cell profile in adipose tissue is associated with insulin resistance and systemic inflammation in humans. *Arteriosclerosis, Thrombosis, and Vascular Biology, 34*(12), 2637–2643.
311. Meaburn, K. J., Burman, B., & Misteli, T. (2016). Spatial genome organization and disease. In *The functional nucleus* (pp. 101–125). Berlin: Springer International Publishing.
312. Meinhardt, H. (1982). *Models of biological pattern formation.* London: Academic.
313. Melmed, S., & Conn, P. M. (Eds.). (2007). Endocrinology: Basic and clinical principles. Berlin: Springer Science & Business Media.
314. Mendes, N. D., Lang, F., Le Cornec, Y. S., Mateescu, R., Batt, G., & Chaouiya, C. (2013). Composition and abstraction of logical regulatory modules: Application to multicellular systems. *Bioinformatics, 29*(6), 749–757.
315. Mendes, P., Hoops, S., Sahle, S., Gauges, R., Dada, J., & Kummer, U. (2009). Computational modeling of biochemical networks using COPASI. In *Methods in Molecular Biology, Systems Biology* (Vol. 500, pp. 17–59). New York: Humana Press.
316. Méndez-López, L. F., Davila-Velderrain, J., Domínguez-Hüttinger, E., Enríquez-Olguín, C., Martínez-García, J. C., & Álvarez-Buylla, E. R. (2017). Gene regulatory network underlying the immortalization of epithelial cells. *BMC Systems Biology, 11*(1), 24.
317. Mendoza, L. (2006). A network model for the control of the differentiation process in Th cells. *Biosystems, 84*(2), 101–114.
318. Mendoza, L., & Álvarez-Buylla, E. R. (1998). Dynamics of the genetic regulatory network for arabidopsis thaliana flower morphogenesis. *Journal of Theoretical Biology, 193*(2), 307–319.
319. Mendoza, L., & Álvarez-Buylla, E. R. (2000). Genetic regulation of root hair development in Arabidopsis thaliana: A network model. *Journal of Theoretical Biology, 204*, 311–326.
320. Mendoza, L., Thieffry, D., & Álvarez-Buylla, E. R. (1999). Genetic control of flower morphogenesis in Arabidopsis thaliana: A logical analysis. *Bioinformatics, 15*, 593–606.
321. Meng, L., Maskarinec, G., Lee, J., & Kolonel, L. N. (1999). Lifestyle factors and chronic diseases: Application of a composite risk index. *Preventive Medicine, 29*(4), 296–304.
322. Micalizzi, D. S., Farabaugh, S. M., & Ford, H. L. (2010). Epithelial-mesenchymal transition in cancer: Parallels between normal development and tumor progression. *Journal of Mammary Gland Biology and Neoplasia, 15*(2), 117–134.
323. Milanovic, M., Fan, D. N., Belenki, D., Däbritz, J. H. M., Zhao, Z., Yu, Y., et al. (2018). Senescence-associated reprogramming promotes cancer stemness. *Nature, 553*, 96–100.
324. Milo, R. (2013). What is the total number of protein molecules per cell volume? A call to rethink some published values. *Bioessays, 35*(12), 1050–1055.
325. Milo, R., Shen-Orr, S., Itzkovitz, S., Kashtan, N., Chklovskii, D., & Alon, U. (2002). Network motifs: Simple building blocks of complex networks. *Science, 298*(5594), 824–827.
326. Mirabet, V., Das, P., Boudaoud, A., & Hamant, O. (2011). The role of mechanical forces in plant morphogenesis. *Annual Review of Plant Biology, 62*, 365–385.

327. Molina, D. M., Jafari, R., Ignatushchenko, M., Seki, T., Larsson, E. A., Dan, C., et al. (2013). Monitoring drug target engagement in cells and tissues using the cellular thermal shift assay. *Science, 341*(6141), 84–87.

328. Monk, N. A. (2003). Oscillatory expression of Hes1, p53, and NF-$\kappa$B driven by transcriptional time delays. *Current Biology, 13*(16), 1409–1413.

329. Morel, A. P., Lièvre, M., Thomas, C., Hinkal, G., Ansieau, S., & Puisieux, A. (2008). Generation of breast cancer stem cells through epithelial-mesenchymal transition. *PloS One, 3*(8), e2888.

330. Moris, N., Pina, C., & Arias, A. M. (2016). Transition states and cell fate decisions in epigenetic landscapes. *Nature Reviews Genetics, 17*(11), 693–703.

331. Mosmann, T. R., Cherwinski, H., Bond, M. W., Giedlin, M. A., & Coffman, R. L. (1986). Two types of murine helper T cell clone. I. Definition according to profiles of lymphokine activities and secreted proteins. *The Journal of Immunology, 136*(7), 2348–2357.

332. Müller, J., & Tjardes, T. (2003). Modeling the cytokine network in vitro and in vivo. *Computational and Mathematical Methods in Medicine, 5*(2), 93–110.

333. Muñoz-Espín, D., Cañamero, M., Maraver, A., Gómez-López, G., Contreras, J., Murillo-Cuesta, S., et al. (2013). Programmed cell senescence during mammalian embryonic development. *Cell, 155*(5), 1104–1118.

334. Murphy, K. M., & Stockinger, B. (2010). Effector T cell plasticity: Flexibility in the face of changing circumstances. *Nature Immunology, 11*(8), 674–680.

335. Murray, J. D. (2003). *Mathematical biology: II: Spatial models and biomedical applications. Interdisciplinary applied mathematics* (3rd ed., Vol. 18). New York: Springer.

336. Müssel, C., Hopfensitz, M., & Kestler, H. A. (2010). BoolNet–An R package for generation, reconstruction and analysis of Boolean networks. *Bioinformatics, 26*(10), 1378–1380.

337. Myers, J. M. B., Martin, L. J., Kovacic, M. B., Mersha, T. B., He, H., Pilipenko, V., et al. (2014). Epistasis between serine protease inhibitor Kazal-type 5 (SPINK5) and thymic stromal lymphopoietin (TSLP) genes contributes to childhood asthma. *Journal of Allergy and Clinical Immunology, 134*(4), 891–899.

338. Nakamura, Y., Oscherwitz, J., Cease, K. B., Chan, S. M., Muñoz-Planillo, R., Hasegawa, M., et al. (2013). Staphylococcus $\delta$-toxin induces allergic skin disease by activating mast cells. *Nature, 503*(7476), 397–401.

339. Nakayamada, S., Takahashi, H., Kanno, Y., & O'Shea, J. J. (2012). Helper T cell diversity and plasticity. *Current Opinion in Immunology, 24*(3), 297–302.

340. Naldi, A., Berenguier, D., Fauré, A., Lopez, F., Thieffry, D., & Chaouiya, C. (2009). Logical modelling of regulatory networks with GINsim 2.3. *Biosystems, 97*(2), 134–139.

341. Naldi, A., Carneiro, J., Chaouiya, C., & Thieffry, D. (2010). Diversity and plasticity of Th cell types predicted from regulatory network modelling. *PLoS Computational Biology, 6*(9), e1000912.

342. Naldi, A., Monteiro, P. T., Müssel, C., Consortium for Logical Models and Tools, Kestler, H. A., Thieffry, D., et al. (2015). Cooperative development of logical modelling standards and tools with CoLoMoTo. *Bioinformatics, 31*(7), 1154–1159.

343. Nam, J., Aguda, B. D., Rath, B., & Agarwal, S. (2009). Biomechanical thresholds regulate inflammation through the NF-$\kappa$B pathway: Experiments and modeling. *PLoS One, 4*(4), e5262.

344. Naryzhny, S. N., Lisitsa, A. V., Zgoda, V. G., Ponomarenko, E. A., & Archakov, A. I. (2014). 2DE-based approach for estimation of number of protein species in a cell. *Electrophoresis, 35*(6), 895–900.

345. Nestle, F. O., Di Meglio, P., Qin, J. Z., & Nickoloff, B. J. (2009). Skin immune sentinels in health and disease. *Nature Reviews Immunology, 9*(10), 679–691.

346. Nevozhay, D., Adams, R. M., Murphy, K. F., Josić, K., & Balázsi, G. (2009). Negative autoregulation linearizes the dose-response and suppresses the heterogeneity of gene expression. *Proceedings of the National Academy of Sciences, 106*(13), 5123–5128.

347. Newell, L., Polak, M. E., Perera, J., Owen, C., Boyd, P., Pickard, C., et al. (2013). Sensitization via healthy skin programs Th2 responses in individuals with atopic dermatitis. *Journal of Investigative Dermatology, 133*(10), 2372–2380.

348. Nieto, M. A., Huang, R. Y. J., Jackson, R. A., & Thiery, J. P. (2016). EMT: 2016. *Cell, 166*(1), 21–45.

349. Nnoruka, E. N., & Dermatology London Clinic. (2004). Tropical medicine rounds current epidemiology of atopic dermatitis in south-eastern Nigeria *International Journal of Dermatology, 43*(10), 739–744. Wiley online library, ISSN:1365-4632.

350. Nograles, K. E., Zaba, L. C., Guttman-Yassky, E., Fuentes-Duculan, J., Suárez-Fariñas, M., Cardinale, I., et al. (2008). Th17 cytokines interleukin (IL)-17 and IL-22 modulate distinct inflammatory and keratinocyte-response pathways. *British Journal of Dermatology, 159*(5), 1092–1102.

351. Norris, D. A. (2005). Mechanisms of action of topical therapies and the rationale for combination therapy. *Journal of the American Academy of Dermatology, 53*(1), S17–S25.

352. Nuorti, J. P., Butler, J. C., Crutcher, J. M., Guevara, R., Welch, D., Holder, P., et al. (1998). An outbreak of multidrug-resistant pneumococcal pneumonia and bacteremia among unvaccinated nursing home residents. *New England Journal of Medicine, 338*(26), 1861–1868.

353. Oates, A. C. (2011). What's all the noise about developmental stochasticity?. *Development, 138*(4), 601–607.

354. Olefsky, J. M., & Glass, C. K. (2010). Macrophages, inflammation, and insulin resistance. *Annual Review of Physiology, 72*, 219–246.

355. Olsen, L., Sherratt, J. A., & Maini, P. K. (1996). A mathematical model for fibro-proliferative wound healing disorders. *Bulletin of Mathematical Biology, 58*(4), 787–808.

356. O'regan, G. M., Sandilands, A., McLean, W. I., & Irvine, A. D. (2009). Filaggrin in atopic dermatitis. *Journal of Allergy and Clinical Immunology, 124*(3), R2–R6.

357. O'Shea, J. J., Gadina, M., & Kanno, Y. (2011). Cytokine signaling: Birth of a pathway. *The Journal of Immunology, 187*(11), 5475–5478.

358. O'Shea, J. J., & Paul, W. E. (2010). Mechanisms underlying lineage commitment and plasticity of helper CD4+ T cells. *Science, 327*(5969), 1098–1102.

359. Oyarzún, D. A., Chaves, M., & Hoff-Hoffmeyer-Zlotnik, M. (2012). Multistability and oscillations in genetic control of metabolism. *Journal of Theoretical Biology, 295*, 139–153.

360. Ozbudak, E. M., Thattai, M., Lim, H. N., Shraiman, B. I., & Van Oudenaarden, A. (2004). Multistability in the lactose utilization network of Escherichia coli. *Nature, 427*(6976), 737–740.

361. Palmer, C. N., Irvine, A. D., Terron-Kwiatkowski, A., Zhao, Y., Liao, H., Lee, S. P., et al. (2006). Common loss-of-function variants of the epidermal barrier protein filaggrin are a major predisposing factor for atopic dermatitis. *Nature Genetics, 38*(4), 441–446.

362. Pantoja-Hernández, L., & Martínez-García, J. C. (2015). Retroactivity in the context of modularly structured biomolecular systems. *Frontiers in Bioengineering and Biotechnology, 3*, 85.

363. Park, S., & Lehner, B. (2015). Cancer type-dependent genetic interactions between cancer driver alterations indicate plasticity of epistasis across cell types. *Molecular Systems Biology, 11*(7), 824.

364. Paternoster, L., Standl, M., Waage, J., Baurecht, H., Hotze, M., Strachan, D. P., et al. (2015). Multi-ancestry genome-wide association study of 21,000 cases and 95,000 controls identifies new risk loci for atopic dermatitis. *Nature Genetics, 47*(12), 1449–1456.

365. Paulsen, M., Legewie, S., Eils, R., Karaulanov, E., & Niehrs, C. (2011). Negative feedback in the bone morphogenetic protein 4 (BMP4) synexpression group governs its dynamic signaling range and canalizes development. *Proceedings of the National Academy of Sciences, 108*(25), 10202–10207.

366. Pawankar, R. (2014). Allergic diseases and asthma: A global public health concern and a call to action. *World Allergy Organization Journal, 7*(1), 12.

367. Pearson, K. J., Baur, J. A., Lewis, K. N., Peshkin, L., Price, N. L., Labinskyy, N., et al. (2008). Resveratrol delays age-related deterioration and mimics transcriptional aspects of dietary restriction without extending life span. *Cell Metabolism, 8*(2), 157–168.

368. Pedersen, E., & Bongo, L. A. (2017). Large-scale biological meta-database management. *Future Generation Computer Systems, 67,* 481–489.

369. Pérez-Ruiz, R. V., García-Ponce, B., Marsch-Martínez, N., Ugartechea-Chirino, Y., Villajuana-Bonequi, M., de Folter, S., et al. (2015). XAANTAL2 (AGL14) is an important component of the complex gene regulatory network that underlies Arabidopsis shoot apical meristem transitions. *Molecular Plant, 8*(5), 796–813.

370. Perkins, T. J., & Swain, P. S. (2009). Strategies for cellular decision-making. *Molecular Systems Biology, 5*(1), 326.

371. Pfeuty, B., & Kaneko, K. (2014). Reliable binary cell-fate decisions based on oscillations. *Physical Review E, 89*(2), 022707.

372. Pisco, A. O., & Huang, S. (2015). Non-genetic cancer cell plasticity and therapy-induced stemness in tumour relapse: 'What does not kill me strengthens me'. *British Journal of Cancer, 112*(11), 1725.

373. Plsek, P., & Greenhalgh, T. (2001). The challenge of complexity in health care: An introduction. *BMJ, 323*(7314), 625–628.

374. Podtschaske, M., Benary, U., Zwinger, S., Höfer, T., Radbruch, A., & Baumgrass, R. (2007). Digital NFATc2 activation per cell transforms graded T cell receptor activation into an all-or-none IL-2 expression. *PLoS One, 2*(9), e935.

375. Polak, P., Karlić, R., Koren, A., Thurman, R., Sandstrom, R., Lawrence, M. S., et al. (2015). Cell-of-origin chromatin organization shapes the mutational landscape of cancer. *Nature, 518*(7539), 360–364.

376. Poplawski, N. J., Swat, M., Gens, J. S., & Glazier, J. A. (2007). Adhesion between cells, diffusion of growth factors, and elasticity of the AER produce the paddle shape of the chick limb. *Physica A: Statistical Mechanics and Its Applications, 373,* 521–532.

377. Prina, E., Ranzani, O. T., & Torres, A. (2015). Community-acquired pneumonia. *The Lancet, 386*(9998), 1097–1108.

378. Prochiantz, A., & Joliot, A. (2003). Can transcription factors function as cell–cell signalling molecules?. *Nature Reviews Molecular Cell Biology, 4*(10), 814–819.

379. Pujadas, E., & Feinberg, A. P. (2012). Regulated noise in the epigenetic landscape of development and disease. *Cell, 148*(6), 1123–1131.

380. Purvis, J. E., Karhohs, K. W., Mock, C., Batchelor, E., Loewer, A., & Lahav, G. (2012). p53 dynamics control cell fate. *Science, 336*(6087), 1440–1444.

381. Purvis, J. E., & Lahav, G. (2013). Encoding and decoding cellular information through signaling dynamics. *Cell, 152*(5), 945–956.

382. Quail, D. F., Olson, O. C., Bhardwaj, P., Walsh, L. A., Akkari, L., Quick, M. L., et al. (2017). Obesity alters the lung myeloid cell landscape to enhance breast cancer metastasis through IL5 and GM-CSF. *Nature Cell Biology, 19*(8), ncb3578.

383. Radisky, D. C., & LaBarge, M. A. (2008). Epithelial-mesenchymal transition and the stem cell phenotype. *Cell Stem Cell, 2*(6), 511–512.

384. Raviv, S., Bharti, K., Rencus-Lazar, S., Cohen-Tayar, Y., Schyr, R., Evantal, N., et al. (2014). PAX6 regulates melanogenesis in the retinal pigmented epithelium through feed-forward regulatory interactions with MITF. *PLoS Genetics, 10*(5), e1004360.

385. Richard, C. Y., Pesce, C. G., Colman-Lerner, A., Lok, L., Pincus, D., Serra, E., et al. (2008). Negative feedback that improves information transmission in yeast signalling. *Nature, 456*(7223), 755–761.

386. Richard, M., & Yvert, G. (2014). How does evolution tune biological noise?. *Frontiers in Genetics, 5,* 374.

387. Ring, J., Alomar, A., Bieber, T., Deleuran, M., Fink-Wagner, A., Gelmetti, C., et al. (2012). Guidelines for treatment of atopic eczema (atopic dermatitis) Part I. *Journal of the European Academy of Dermatology and Venereology, 26*(8), 1045–1060.

388. Rivera, C. M., & Ren, B. (2013). Mapping human epigenomes. *Cell, 155*(1), 39–55.

389. Roedl, D., Traidl-Hoffmann, C., Ring, J., Behrendt, H., & Braun-Falco, M. (2009). Serine protease inhibitor lymphoepithelial Kazal type-related inhibitor tends to be decreased in atopic dermatitis. *Journal of the European Academy of Dermatology and Venereology, 23*(11), 1263.

390. Roekevisch, E., Spuls, P. I., Kuester, D., Limpens, J., & Schmitt, J. (2014). Efficacy and safety of systemic treatments for moderate-to-severe atopic dermatitis: A systematic review. *Journal of Allergy and Clinical Immunology, 133*(2), 429–438.

391. Rogers, E. D., Jackson, T., Moussaieff, A., Aharoni, A., & Benfey, P. N. (2012). Cell type-specific transcriptional profiling: Implications for metabolite profiling. *The Plant Journal, 70*(1), 5–17.

392. Romero, I. G., Ruvinsky, I., & Gilad, Y. (2012). Comparative studies of gene expression and the evolution of gene regulation. *Nature Reviews Genetics, 13*(7), 505–516.

393. Rubin, H. (1985). Cancer as a dynamic developmental disorder. *Cancer Research, 45*(7), 2935–2942.

394. Rybinski, B., Franco-Barraza, J., & Cukierman, E. (2014). The wound healing, chronic fibrosis, and cancer progression triad. *Physiological Genomics, 46*(7), 223–244.

395. Ryu, H., Chung, M., Dobrzyński, M., Fey, D., Blum, Y., Lee, S. S., et al. (2015). Frequency modulation of ERK activation dynamics rewires cell fate. *Molecular Systems Biology, 11*(11), 838.

396. Sakaguchi, S., Sakaguchi, N., Asano, M., Itoh, M., & Toda, M. (1995). Immunologic self-tolerance maintained by activated T cells expressing IL-2 receptor alpha-chains (CD25). Breakdown of a single mechanism of self-tolerance causes various autoimmune diseases. *The Journal of Immunology, 155*(3), 1151–1164.

397. Sandilands, A., Sutherland, C., Irvine, A. D., & McLean, W. I. (2009). Filaggrin in the frontline: Role in skin barrier function and disease. *Journal of Cell Science, 122*(9), 1285–1294.

398. Sandmann, T., Girardot, C., Brehme, M., Tongprasit, W., Stolc, V., & Furlong, E. E. (2007). A core transcriptional network for early mesoderm development in Drosophila melanogaster. *Genes & Development, 21*(4), 436–449.

399. Saraiva, M., & O'garra, A. (2010). The regulation of IL-10 production by immune cells. *Nature Reviews Immunology, 10*(3), 170–181.

400. Savignac, M., Edir, A., Simon, M., & Hovnanian, A. (2011). Darier disease: A disease model of impaired calcium homeostasis in the skin. *Biochimica et Biophysica Acta—Molecular Cell Research, 1813*(5), 1111–1117.

401. Sawyer, J. M., Harrell, J. R., Shemer, G., Sullivan-Brown, J., Roh-Johnson, M., & Goldstein, B. (2010). Apical constriction: A cell shape change that can drive morphogenesis. *Developmental Biology, 341*(1), 5–19.

402. Schaerli, P., Willimann, K., Lang, A. B., Lipp, M., Loetscher, P., & Moser, B. (2000). CXC chemokine receptor 5 expression defines follicular homing T cells with B cell helper function. *Journal of Experimental Medicine, 192*(11), 1553–1562.

403. Scharschmidt, T. C., Man, M. Q., Hatano, Y., Crumrine, D., Gunathilake, R., Sundberg, J. P., et al. (2009). Filaggrin deficiency confers a paracellular barrier abnormality that reduces inflammatory thresholds to irritants and haptens. *Journal of Allergy and Clinical Immunology, 124*(3), 496–506.

404. Schilstra, M. J., & Nehaniv, C. L. (2008). Bio-logic: Gene expression and the laws of combinatorial logic. *Artificial Life, 14*(1), 121–133.

405. Schmidt, H., & Jirstrand, M. (2005). Systems biology toolbox for MATLAB: A computational platform for research in systems biology. *Bioinformatics, 22*(4), 514–515.

406. Schmitt, E., Klein, M., & Bopp, T. (2014). Th9 cells, new players in adaptive immunity. *Trends in Immunology, 35*(2), 61–68.

407. Schoepe, S., Schäcke, H., May, E., & Asadullah, K. (2006). Glucocorticoid therapy-induced skin atrophy. *Experimental Dermatology, 15*(6), 406–420.

408. Schrag, S. J., Peña, C., Fernández, J., Sánchez, J., Gómez, V., Pérez, E., et al. (2001). Effect of short-course, high-dose amoxicillin therapy on resistant pneumococcal carriage: A randomized trial. *JAMA, 286*(1), 49–56.

409. Schwabe, R. F., & Jobin, C. (2013). The microbiome and cancer. *Nature Reviews Cancer, 13*(11), 800–812.

410. Setty, Y., Mayo, A. E., Surette, M. G., & Alon, U. (2003). Detailed map of a cis-regulatory input function. *Proceedings of the National Academy of Sciences, 100*(13), 7702–7707.

411. Sewell, G. W., Marks, D. J., & Segal, A. W. (2009). The immunopathogenesis of Crohn's disease: A three-stage model. *Current Opinion in Immunology, 21*(5), 506–513.

412. Shah, N. A., & Sarkar, C. A. (2011). Robust network topologies for generating switch-like cellular responses. *PLoS Computational Biology, 7*(6), e1002085.

413. Shalek, A. K., Satija, R., Adiconis, X., Gertner, R. S., Gaublomme, J. T., Raychowdhury, R., et al. (2013). Single-cell transcriptomics reveals bimodality in expression and splicing in immune cells. *Nature, 498*(7453), 236–240.

414. Sharlow, E. R., Paine, C. S., Babiarz, L., Eisinger, M., Shapiro, S., & Seiberg, M. (2000). The protease-activated receptor-2 upregulates keratinocyte phagocytosis. *Journal of Cell Science, 113*(17), 3093–3101.

415. Shih, C., Padhy, L. C., Murray, M., & Weinberg, R. A. (1981). Transforming genes of carcinomas and neuroblastomas introduced into mouse fibroblasts. *Nature, 290*, 261–264.

416. Shih, C., & Weinberg, R. A. (1982). Isolation of a transforming sequence from a human bladder carcinoma cell line. *Cell, 29*(1), 161–169.

417. Shin, S., Seong, J. K., & Bae, Y. S. (2016). Ahnak stimulates BMP2-mediated adipocyte differentiation through Smad1 activation. *Obesity, 24*(2), 398–407.

418. Sidbury, R., Davis, D. M., Cohen, D. E., Cordoro, K. M., Berger, T. G., Bergman, J. N., et al. (2014). Guidelines of care for the management of atopic dermatitis: Section 3. Management and treatment with phototherapy and systemic agents. *Journal of the American Academy of Dermatology, 71*(2), 327–349.

419. Simões-Costa, M., & Bronner, M. E. (2015). Establishing neural crest identity: A gene regulatory recipe. *Development, 142*(2), 242–257.

420. Simpson, E. L., Chalmers, J. R., Hanifin, J. M., Thomas, K. S., Cork, M. J., McLean, W. I., et al. (2014). Emollient enhancement of the skin barrier from birth offers effective atopic dermatitis prevention. *Journal of Allergy and Clinical Immunology, 134*(4), 818–823.

421. Soetaert, K. E. R., Petzoldt, T., & Setzer, R. W. (2010). Solving differential equations in R: Package deSolve. *Journal of Statistical Software, 33*, 1–25. ISSN: 1548-7660.

422. Solé, R. V., & Goodwin, B. C. (2000). *Signs of life: How complexity pervades biology.* London: Basic Books.

423. Sotiropoulou, G., & Pampalakis, G. (2010). Kallikrein-related peptidases: Bridges between immune functions and extracellular matrix degradation. *Biological Chemistry, 391*(4), 321–331.

424. Soto, A. M., & Sonnenschein, C. (2004). The somatic mutation theory of cancer: Growing problems with the paradigm?. *Bioessays, 26*(10), 1097–1107.

425. Steinway, S. N., Zañudo, J. G. T., Michel, P. J., Feith, D. J., Loughran, T. P., & Albert, R. (2015). Combinatorial interventions inhibit TGF $\beta$-driven epithelial-to-mesenchymal transition and support hybrid cellular phenotypes. *NPJ Systems Biology and Applications, 1*, 15014.

426. Stelling, J., Sauer, U., Szallasi, Z., Doyle, F. J., & Doyle, J. (2004). Robustness of cellular functions. *Cell, 118*(6), 675–685.

427. Storer, M., Mas, A., Robert-Moreno, A., Pecoraro, M., Ortells, M. C., Di Giacomo, V., et al. (2013). Senescence is a developmental mechanism that contributes to embryonic growth and patterning. *Cell, 155*(5), 1119–1130.

428. Stratton, M. R., Campbell, P. J., & Futreal, P. A. (2009). The cancer genome. *Nature, 458*(7239), 719–724.

429. Strogatz, S. H. (2014). *Nonlinear dynamics and chaos with applications to physics, biology, chemistry, and engineering* (2nd ed.). Boulder: Westview Press.

430. Subramanian, A., Tamayo, P., Mootha, V. K., Mukherjee, S., Ebert, B. L., Gillette, M. A., et al. (2005). Gene set enrichment analysis: A knowledge-based approach for interpreting genome-wide expression profiles. *Proceedings of the National Academy of Sciences, 102*(43), 15545–15550.
431. Sugiura, A., Nomura, T., Mizuno, A., & Imokawa, G. (2014). Reevaluation of the non-lesional dry skin in atopic dermatitis by acute barrier disruption: An abnormal permeability barrier homeostasis with defective processing to generate ceramide. *Archives of Dermatological Research, 306*(5), 427–440.
432. Sung, M. H., Li, N., Lao, Q., Gottschalk, R. A., Hager, G. L., & Fraser, I. D. (2014). Switching of the relative dominance between feedback mechanisms in lipopolysaccharide-induced NF-$\kappa$B signaling. *Science Signaling, 7*(308), ra6.
433. Sütterlin, T., Huber, S., Dickhaus, H., & Grabe, N. (2009). Modeling multi-cellular behavior in epidermal tissue homeostasis via finite state machines in multi-agent systems. *Bioinformatics, 25*(16), 2057–2063.
434. Sütterlin, T., Kolb, C., Dickhaus, H., Jäger, D., & Grabe, N. (2012). Bridging the scales: Semantic integration of quantitative SBML in graphical multi-cellular models and simulations with EPISIM and COPASI. *Bioinformatics, 29*(2), 223–229.
435. Szallasi, Z., Periwal, V., & Stelling, J. (2006). *System modeling in cellular biology: From concepts to nuts and bolts*. Cambridge: The MIT Press.
436. Taïeb, A., Seneschal, J., & Mossalayi, M. D. (2012). Biologics in atopic dermatitis. *JDDG: Journal der Deutschen Dermatologischen Gesellschaft, 10*(3), 174–178.
437. Takai, T., & Ikeda, S. (2011). Barrier dysfunction caused by environmental proteases in the pathogenesis of allergic diseases. *Allergology International, 60*(1), 25–35.
438. Tanaka, G., Christodoulides, P., Domínguez-Hüttinger, E., Aihara, K., & Tanaka, R. J. (2018). Bifurcation analysis of a mathematical model of atopic dermatitis to determine patient-specific effects of treatments on dynamic phenotypes. *Journal of Theoretical Biology, 448*, 66–79.
439. Tanaka, R. J., Ono, M., & Harrington, H. A. (2011). Skin barrier homeostasis in atopic dermatitis: Feedback regulation of kallikrein activity. *PloS One, 6*(5), e19895.
440. Tanay, A., & Regev, A. (2017). Scaling single-cell genomics from phenomenology to mechanism. *Nature, 541*(7637), 331–338.
441. Tang, T. S., Bieber, T., & Williams, H. C. (2014). Are the concepts of induction of remission and treatment of subclinical inflammation in atopic dermatitis clinically useful?. *Journal of Allergy and Clinical Immunology, 133*(6), 1615–1625.
442. Tawfik, D. S. (2010). Messy biology and the origins of evolutionary innovations. *Nature Chemical Biology, 6*(10), 692.
443. Tay, S., Hughey, J. J., Lee, T. K., Lipniacki, T., Quake, S. R., & Covert, M. W. (2010). Single-cell NF-$\kappa$B dynamics reveal digital activation and analogue information processing. *Nature, 466*(7303), 267–271.
444. Terui, T., Hirao, T., Sato, Y., Uesugi, T., Honda, M., Iguchi, M., et al. (1998). An increased ratio of interleukin-1 receptor antagonist to interleukin-1$\alpha$ in inflammatory skin diseases. *Experimental Dermatology, 7*(6), 327–334.
445. Theodorakis, C. W. (2001). Integration of genotoxic and population genetic endpoints in biomonitoring and risk assessment. *Ecotoxicology, 10*(4), 245–256.
446. Thom, R. (1983). *Paraboles et catastrophes*. Paris: Flammarion.
447. Thorén, H., & Gerlee, P. (2010). Weak emergence and complexity. In *Artificial Life XII Proceedings of the Twelfth International Conference on the Synthesis and Simulation of Living Systems* (pp. 879–886). Cambridge: MIT Press.
448. Thyssen, J. P., & Kezic, S. (2014). Causes of epidermal filaggrin reduction and their role in the pathogenesis of atopic dermatitis. *Journal of Allergy and Clinical Immunology, 134*(4), 792–799.
449. Tiemann, C. A., Vanlier, J., Hilbers, P. A., & van Riel, N. A. (2011). Parameter adaptations during phenotype transitions in progressive diseases. *BMC Systems Biology, 5*(1), 174.

450. Tiemann, C. A., Vanlier, J., Oosterveer, M. H., Groen, A. K., Hilbers, P. A., & van Riel, N. A. (2013). Parameter trajectory analysis to identify treatment effects of pharmacological interventions. *PLoS Computational Biology, 9*(8), e1003166.

451. Tiwari, A., Ray, J. C. J., Narula, J., & Igoshin, O. A. (2011). Bistable responses in bacterial genetic networks: Designs and dynamical consequences. *Mathematical Biosciences, 231*(1), 76–89.

452. Tomlin, C. J., & Axelrod, J. D. (2007). Biology by numbers: Mathematical modelling in developmental biology. *Nature Reviews Genetics, 8*(5), 331.

453. Toni, T., & Stumpf, M. P. (2009). Simulation-based model selection for dynamical systems in systems and population biology. *Bioinformatics, 26*(1), 104–110.

454. Transtrum, M. K., & Qiu, P. (2016). Bridging mechanistic and phenomenological models of complex biological systems. *PLoS Computational Biology, 12*(5), e1004915.

455. Tsai, C. C., Chen, Y. J., Yew, T. L., Chen, L. L., Wang, J. Y., Chiu, C. H., & Hung, S. C. (2011). Hypoxia inhibits senescence and maintains mesenchymal stem cell properties through down-regulation of E2A-p21 by HIF-TWIST. *Blood, 117*(2), 459–469.

456. Tu, C. L., Crumrine, D. A., Man, M. Q., Chang, W., Elalieh, H., You, M., et al. (2012). Ablation of the calcium-sensing receptor in keratinocytes impairs epidermal differentiation and barrier function. *Journal of Investigative Dermatology, 132*(10), 2350–2359.

457. Turing, A. M. (1952). The chemical basis of morphogenesis. *Philosophical Transactions of the Royal Society of London B, 237*(641), 37–72.

458. Turner, J. R. (2009). Intestinal mucosal barrier function in health and disease. *Nature Reviews Immunology, 9*(11), 799–809.

459. Tyson, J. J., Chen, K. C., & Novak, B. (2003). Sniffers, buzzers, toggles and blinkers: Dynamics of regulatory and signaling pathways in the cell. *Current Opinion in Cell Biology, 15*(2), 221–231.

460. USDA-NRCS PLANTS Database/Britton, N. L., & Brown, A. (1913). *An illustrated flora of the Northern United States, Canada and the British Possessions* (3 vols., Vol. 2, p. 176). New York: Charles Scribner's Sons.

461. Valeyev, N. V., Hundhausen, C., Umezawa, Y., Kotov, N. V., Williams, G., Clop, A., et al. (2010). A systems model for immune cell interactions unravels the mechanism of inflammation in human skin. *PLoS Computational Biology, 6*(12), e1001024.

462. Van Kampen, N. G. (1992). *Stochastic processes in physics and chemistry* (revised and enlarged edition). Amsterdam: Elsevier.

463. Van Kruijsdijk, R. C., van der Wall, E., & Visseren, F. L. (2009). Obesity and cancer: The role of dysfunctional adipose tissue. *Cancer Epidemiology and Prevention Biomarkers, 18*(10), 2569–2578.

464. van Riel, N. A., Tiemann, C. A., Vanlier, J., & Hilbers, P. A. (2013). Applications of analysis of dynamic adaptations in parameter trajectories. *Interface Focus, 3*(2), 20120084.

465. Veening, J. W., Smits, W. K., & Kuipers, O. P. (2008). Bistability, epigenetics, and bet-hedging in bacteria. *Annual Review of Microbiology, 62*, 193–210.

466. Velderrain, J. D., Martínez-García, J. C., & Álvarez-Buylla, E. R. (2017). Boolean dynamic modeling approaches to study plant gene regulatory networks: Integration, validation, and prediction. In *Methods in molecular biology (Clifton, NJ)*, (Vol. 629, pp. 297–315). New York: Humana Press.

467. Veldhoen, M., Hocking, R. J., Atkins, C. J., Locksley, R. M., & Stockinger, B. (2006). TGFβ in the context of an inflammatory cytokine milieu supports de novo differentiation of IL-17-producing T cells. *Immunity, 24*(2), 179–189.

468. Villarreal, C., Padilla-Longoria, P., & Álvarez-Buylla, E. R. (2012). General theory of genotype to phenotype mapping: Derivation of epigenetic landscapes from N-node complex gene regulatory networks. *Physical Review Letters, 109*(11), 118102.

469. Vinuesa, C. G., Fagarasan, S., & Dong, C. (2013). New territory for T follicular helper cells. *Immunity, 39*(3), 417–420.

470. Virchow, R. (1858). Die Cellularpathologie in ihrer Begründung auf physiologische und pathologische Gewebelehre: zwanzig Vorlesungen, gehalten während der Monate Februar, März und April 1858 im pathologischen Institute zu Berlin. Hirschwald.

471. Vogelstein, B., Fearon, E. R., Hamilton, S. R., Kern, S. E., Preisinger, A. C., Leppert, M., et al. (1988). Genetic alterations during colorectal-tumor development. *New England Journal of Medicine, 319*(9), 525–532.

472. Von Dassow, G., Meir, E., Munro, E. M., & Odell, G. M. (2000). The segment polarity network is a robust developmental module. *Nature, 406*(6792), 188.

473. Waddington, C. H. (1957). *The strategy of the genes. A discussion of some aspects of theoretical biology; with an appendix by H. Kacser.* London: George Allen & Unwin.

474. Wadonda-Kabondo, N., Sterne, J. A. C., Golding, J., Kennedy, C. T. C., Archer, C. B., Dunnill, M. G. S., et al. (2003). A prospective study of the prevalence and incidence of atopic dermatitis in children aged 0–42 months. *British Journal of Dermatology, 149*(5), 1023–1028. Epidemiology and Health Services Research.

475. Wagner, A. (2013). *Robustness and evolvability in living systems.* Princeton: Princeton University Press.

476. Wang, G., Yuan, R., Zhu, X., & Ao, P. (2018). Endogenous molecular-cellular network cancer theory: A systems biology approach. In M. Bizzarri (Ed.), *Systems biology. Methods in molecular biology* (Vol. 1702). New York: Humana Press.

477. Wang, G., Zhu, X., Gu, J., & Ao, P. (2014). Quantitative implementation of the endogenous molecular–cellular network hypothesis in hepatocellular carcinoma. *Interface Focus, 4*(3), 20130064.

478. Wang, R. S., Saadatpour, A., & Albert, R. (2012). Boolean modeling in systems biology: An overview of methodology and applications. *Physical Biology, 9*(5), 055001.

479. Wang, X., Gulbahce, N., & Yu, H. (2011). Network-based methods for human disease gene prediction. *Briefings in Functional Genomics, 10*(5), 280–293.

480. Warrell, M. J. (2011). Intradermal rabies vaccination: The evolution and future of pre-and post-exposure prophylaxis. In *Intradermal immunization* (pp. 139–157). Berlin: Springer.

481. Watson, I. R., Takahashi, K., Futreal, P. A., & Chin, L. (2013). Emerging patterns of somatic mutations in cancer. *Nature Reviews Genetics, 14*(10), 703–718.

482. Weber, M., & Buceta, J. (2013). Dynamics of the quorum sensing switch: Stochastic and non-stationary effects. *BMC Systems Biology, 7*(1), 6.

483. Weidinger, S., Baurecht, H., Wagenpfeil, S., Henderson, J., Novak, N., Sandilands, A., et al. (2008). Analysis of the individual and aggregate genetic contributions of previously identified serine peptidase inhibitor Kazal type 5 (SPINK5), kallikrein-related peptidase 7 (KLK7), and filaggrin (FLG) polymorphisms to eczema risk. *Journal of Allergy and Clinical Immunology, 122*(3), 560–568.

484. Weinberg, R. A. (1991). Tumor suppressor genes. *Science, 254*(5035), 1138–1146.

485. Weiner, H. L., da Cunha, A. P., Quintana, F., & Wu, H. (2011). Oral tolerance. *Immunological Reviews, 241*(1), 241–259.

486. Weinstein, J. N., Collisson, E. A., Mills, G. B., Shaw, K. R. M., Ozenberger, B. A., Ellrott, K., et al. (2013). The cancer genome atlas pan-cancer analysis project. *Nature Genetics, 45*(10), 1113–1120.

487. Weisburger, J. H. (2002). Lifestyle, health and disease prevention: The underlying mechanisms. *European Journal of Cancer Prevention: The Official Journal of the European Cancer Prevention Organization (ECP), 11*, S1-7.

488. Weiss, J. N. (1997). The Hill equation revisited: Uses and misuses. *The FASEB Journal, 11*(11), 835–841.

489. Wilhelm, T. (2009). The smallest chemical reaction system with bistability. *BMC Systems Biology, 3*(1), 90.

490. Willhauck, M. J., Mirancea, N., Vosseler, S., Pavesio, A., Boukamp, P., Mueller, M. M., et al. (2007). Reversion of tumor phenotype in surface transplants of skin SCC cells by scaffold-induced stroma modulation. *Carcinogenesis, 28*(3), 595–610.

491. Williams, J. W., Cui, X., Levchenko, A., & Stevens, A. M. (2008). Robust and sensitive control of a quorum-sensing circuit by two interlocked feedback loops. *Molecular Systems Biology, 4*(1), 234.

492. Wills-Karp, M., Santeliz, J., & Karp, C. L. (2001). The germless theory of allergic disease: Revisiting the hygiene hypothesis. *Nature Reviews Immunology, 1*(1), 69–75.
493. Wolkenhauer, O., Ullah, M., Wellstead, P., & Cho, K. H. (2005). The dynamic systems approach to control and regulation of intracellular networks. *FEBS Letters, 579*(8), 1846–1853.
494. Wolpert, L. (1969). Positional information and the spatial pattern of cellular differentiation. *Journal of Theoretical Biology, 25*(1), 1–47.
495. Wood, T., Burke, J., & Rieseberg, L. (2005). Parallel genotypic adaptation: When evolution repeats itself. In *Genetics of adaptation* (pp. 157–170). Dordrecht: Springer.
496. Wuensche, A. (1998, January). Genomic regulation modeled as a network with basins of attraction. In *Pacific Symposium on Biocomputing* (Vol. 3, pp. 89–102).
497. Xing, F., Saidou, J., & Watabe, K. (2010). Cancer associated fibroblasts (CAFs) in tumor microenvironment. *Frontiers in Bioscience: A Journal and Virtual Library, 15*, 166.
498. Xu, H., Barnes, G. T., Yang, Q., Tan, G., Yang, D., Chou, C. J., et al. (2003). Chronic inflammation in fat plays a crucial role in the development of obesity-related insulin resistance. *Journal of Clinical Investigation, 112*(12), 1821.
499. Xu, J., Lamouille, S., & Derynck, R. (2009). TGF-$\beta$-induced epithelial to mesenchymal transition. *Cell Research, 19*(2), 156–172.
500. Xu, L., Kitani, A., Fuss, I., & Strober, W. (2007). Cutting edge: Regulatory T cells induce CD4+ CD25- Foxp3-T cells or are self-induced to become Th17 cells in the absence of exogenous TGF-$\beta$. *The Journal of Immunology, 178*(11), 6725–6729.
501. Yaffe, M. B. (2013). The scientific drunk and the lamppost: Massive sequencing efforts in cancer discovery and treatment. *Science Signaling, 6*(269), pe13.
502. Yamane, H., Zhu, J., & Paul, W. E. (2005). Independent roles for IL-2 and GATA-3 in stimulating naive CD4+ T cells to generate a Th2-inducing cytokine environment. *Journal of Experimental Medicine, 202*(6), 793–804.
503. Yates, A., Bergmann, C., Van Hemmen, J. L., Stark, J., & Callard, R. (2000). Cytokine-modulated regulation of helper T cell populations. *Journal of Theoretical Biology, 206*(4), 539–560.
504. Ye, X., & Weinberg, R. A. (2015). Epithelial–mesenchymal plasticity: A central regulator of cancer progression. *Trends in Cell Biology, 25*(11), 675–686.
505. Yi, T. M., Huang, Y., Simon, M. I., & Doyle, J. (2000). Robust perfect adaptation in bacterial chemotaxis through integral feedback control. *Proceedings of the National Academy of Sciences, 97*(9), 4649–4653.
506. Yoshimura, A., Naka, T., & Kubo, M. (2007). SOCS proteins, cytokine signalling and immune regulation. *Nature Reviews Immunology, 7*(6), 454–465.
507. Yoshimura, A., Suzuki, M., Sakaguchi, R., Hanada, T., & Yasukawa, H. (2012). SOCS, inflammation, and autoimmunity. *Frontiers in Immunology, 3*, 20.
508. Yu, C., Wang, F., Jin, C., Wu, X., Chan, W. K., & McKeehan, W. L. (2002). Increased carbon tetrachloride-induced liver injury and fibrosis in FGFR4-deficient mice. *The American Journal of Pathology, 161*(6), 2003–2010.
509. Yu, R., Liu, Q., Liu, J., Wang, Q., & Wang, Y. (2016). Concentrations of organophosphorus pesticides in fresh vegetables and related human health risk assessment in Changchun, Northeast China. *Food Control, 60*, 353–360.
510. Yuan, R., Zhu, X., Radich, J. P., & Ao, P. (2016). From molecular interaction to acute promyelocytic leukemia: Calculating leukemogenesis and remission from endogenous molecular-cellular network. *Scientific Reports, 6*, 24307.
511. Zernicka-Goetz, M., & Huang, S. (2010). Stochasticity versus determinism in development: A false dichotomy?. *Nature Reviews Genetics, 11*(11), 743.
512. Zhang, Q., Bhattacharya, S., Kline, D. E., Crawford, R. B., Conolly, R. B., Thomas, R. S., et al. (2010). Stochastic modeling of B lymphocyte terminal differentiation and its suppression by dioxin. *BMC Systems Biology, 4*(1), 40.
513. Zhang, T., Schmierer, B., & Novák, B. (2011). Cell cycle commitment in budding yeast emerges from the cooperation of multiple bistable switches. *Open Biology, 1*(3), 110009.

514. Zhang, Y. (2013). Cancer embryonic stem cell-like attractors alongside deficiency of regulatory restraints of cell-division and cell-cycle. *Journal of Genetic Syndromes and Gene Therapy, 4*(130), 2.
515. Zhang, Z., Clarke, T. B., & Weiser, J. N. (2009). Cellular effectors mediating Th17-dependent clearance of pneumococcal colonization in mice. *The Journal of Clinical Investigation, 119*(7), 1899.
516. Zhao, Y., Ransom, J. F., Li, A., Vedantham, V., von Drehle, M., Muth, A. N., et al. (2007). Dysregulation of cardiogenesis, cardiac conduction, and cell cycle in mice lacking miRNA-1-2. *Cell, 129*(2), 303–317.
517. Zheng, T., Yu, J., Oh, M. H., & Zhu, Z. (2011). The atopic march: Progression from atopic dermatitis to allergic rhinitis and asthma. *Allergy, Asthma & Immunology Research, 3*(2), 67–73.
518. Zhou, J. X., & Huang, S. (2011). Understanding gene circuits at cell-fate branch points for rational cell reprogramming. *Trends in Genetics, 27*(2), 55–62.
519. Zhou, J. X., Samal, A., d'Hérouël, A. F., Price, N. D., & Huang, S. (2016). Relative stability of network states in Boolean network models of gene regulation in development. *Biosystems, 142*, 15–24.
520. Zhu, J., Cote-Sierra, J., Guo, L., & Paul, W. E. (2003). Stat5 activation plays a critical role in Th2 differentiation. *Immunity, 19*(5), 739–748.
521. Zhu, J., Yamane, H., & Paul, W. E. (2009). Differentiation of effector CD4 T cell populations. *Annual Review of Immunology, 28*, 445–489.
522. Zhu, X., Yuan, R., Hood, L., & Ao, P. (2015). Endogenous molecular-cellular hierarchical modeling of prostate carcinogenesis uncovers robust structure. *Progress in Biophysics and Molecular Biology, 117*(1), 30–42.
523. Zitvogel, L., Pietrocola, F., & Kroemer, G. (2017). Nutrition, inflammation and cancer. *Nature Immunology, 18*(8), 843–850.

# Index

## A

adult life-span, 149
affected tissues
    continuous deterioration of, 4
aging, 138, 141
    cell, 140, 149
    robust outcome of, 140
algorithm
    reduction
        knowledge-based, 149
allergic reactions, 191, 192
allergic rhinitis, 189
allergy, 188, 196
analysis
    robustness, 179
androgen dynamics, 83
animal development, 22
animals, 145
ANTELOPE, 77
antibiotics, 189
antimicrobial peptides, 195
apoptosis, 142
    resistance to, 161
Arabidopsis thaliana, 22, 23, 26, 27
asthma, 6, 189
Atopic Dermatitis, xvi, 6, 12, 19, 29, 133–135,
        137, 188, 192, 208, 209
    advanced, 202
    aggravation of, 202
    biology of, 191
    clinically severe, 202
    disease phenotypes of, 189
    early phases of, 199

effects of emollients on the progression of,
        207
    flares of, 192, 193, 195, 196
    hallmarks of, 193
    mathematical models of, 190
    onset and progression of, 191, 212
    optimal treatments for, 208
    patients suffering, 201
    prevalence of, 189
    prevention of the progression of, 202
    progression of, 9, 203, 206, 207
    severe forms of, 192
    severe symptoms of, 191, 201, 202, 207
attractor
    notion, 25
attractors, 24, 66, 67, 73, 74, 77, 78, 87, 88, 98,
        112, 113, 116, 127, 136, 145, 172
    dynamical, 137
    estimating transition probabilities of, 89
    fixed-point, 76
    local characterization of the properties of,
        87
    multiple, 213
    number and size of, 17
    pre-existent pathological, 157
    predicted, 77, 78
    robust, 157
    sequences, 93
    set of uncovered, 77
    signal-independent transitions among,
        87
    transition probabilities of, 36
    uncovered, 76

M. E. Álvarez-Buylla Roces et al., *Modeling Methods for Medical Systems Biology*,
Advances in Experimental Medicine and Biology 1069,
https://doi.org/10.1007/978-3-319-89354-9

Printed in the United States
By Bookmasters